Advanced Metallic Biomaterials

Madalina-Simona BALTATU

Dumitru-Doru BURDUHOS-NERGIS

Diana Petronela BURDUHOS-NERGIS

Petrica VIZUREANU

Gheorghe Asachi Technical University of Iasi, Faculty of Materials Science & Engineering, Romania

Published by **Materials Research Forum LLC**
Millersville, PA 17551, USA

Published as part of the book series
Materials Research Foundations
Volume 118 (2022)
ISSN 2471-8890 (Print)
ISSN 2471-8904 (Online)

Print ISBN 978-1-64490-176-2
ePDF ISBN 978-1-64490-177-9

Distributed worldwide by

Materials Research Forum LLC
105 Springdale Lane
Millersville, PA 17551
USA
http://www.mrforum.com

Printed in the United States of America
10 9 8 7 6 5 4 3 2 1

Table of Contents

Introduction

Since health is the most important thing, the world continues to pay attention to the research and development of alloys for medical applications. In order to improve the quality of life, it is necessary to improve the classic technology of implant implementation and the biomaterial synthesis technology that implements them. The ultimate goal is to promote a new generation of multifunctional implants with long-term performance.

Each type of biomaterial has advantages and disadvantages but the most important is that it is tolerated by the body for a long time (decades) and therefore must meet the functional requirements according to the medical applications in which they are to be used.

The book is an introduction of metallic biomaterials which discusses the main types of metallic biomaterials such as titanium-based alloys, cobalt-based alloys, stainless steels and biodegradable alloys.

Because the evolution of the use of biomaterials has experienced a rapid development, a wide range of such materials being used but the synthesis and overall assessment of their properties and applications is difficult.

In this context, the book "Advanced metallic biomaterials" wants to be a review of the properties and applications of the main categories of metallic biomaterials, being useful to all specialists in technical and medical fields who want to synthesize and design new materials.

The book contains six chapters.

The first chapter entitled *"General considerations of biomaterials"* presents introductory concepts of biomaterials, clasificassion and a short description of each class of biomaterials.

In the second chapter entitled *"Titanium alloys"* is presented an actual state of the art, of titanium alloys with appropriate characteristics for use in the medical field. The chapter contains the properties of titanium alloys, their classification as well as the applications where titanium-based alloys are found.

The third chapter entitled *"Cobalt alloys"*, contains the properties of cobalt alloys, their classification and influence of elements, as well as the applications where cobalt alloys are found.

Chapter four entitled *"Stainless steel alloys"*, presents the properties of stainless steel allos, influence of alloying elements on stainless steel alloys, classification and aplications in medicine.

The fifth chapter entitled *"Biodegradables alloys"*, also presents a complete characterization of magnesium alloys with description of properties, classification and medical applications.

The sixth chapter entitled *"Optimization of metallic biomaterials"* presents opportunities of variants that could be optimized the metallic biomaterials. Optimization of these types of materials requires a surface functionality and their specialization depends on the followed goal.

CHAPTER 1

General Considerations of Biomaterials

1.1. Introductory notions about biomaterials

Exploration in the field of materials science and designing has extended incredibly in late many years, particularly in the field of biocompatible materials. This is on the grounds that, from one perspective, medication is continually searching for answers to cure numerous medical conditions, and then again, certain classes of materials have effectively demonstrated valuable in enhancing or in any event, relieving specific human misery. The improvement of biocompatible materials research is a developing cycle driven by the increment in the quantity of mishaps and numerous medical issues, yet in addition by the craving to build a normal future for people. As examination in the field of biomaterials advances at the research center level, the occurrence of genuine illnesses in the worldwide human local area is expanding. The proportion of the elderly in the earth's population continues to increase, leading to an increase in cases of chronic diseases, including heart disease and cancer. Infectious diseases are also a major burden on global health; death rates from HIV/AIDS, tuberculosis, infectious diseases and gastrointestinal diseases remain high. Therefore, doctors and researchers in the biomedical field must be prepared to respond to the growing demand for medical care for any types of disease or trauma anywhere in the world. Conventional biomaterials depend on polymers, ceramics and metals, while the most recent age of biomaterials now consolidate biomolecules, helpful medications and surprisingly living cells. As of now, biomaterials are a unique class of materials, irreplaceable for raising the nature of human existence and broadening its span. Biomaterials are by and large expected to be embedded in a living organic entity to reestablish the shape and capacity of a piece of a tissue annihilated by sickness or injury. Before discussing the unique and fundamental potential of biomaterials in disease prevention and treatment, it is important to define the ideal biomaterial and then describe the scope of biomaterial science [1-3].

The definition given to biomaterials has changed permanently in relation to the knowledge acquired over time by specialists in the field and with the increase in the performance of materials obtained by various methods [4-5].

Biomaterials are any substance or combination of substances, of natural or synthetic origin, which can be used for a specified period of time, as a whole or as a component part of a system which treats or replaces a tissue, organ or function. of the human body (1992). A biomaterial is a non-viable (non-living) material used as a medical device to interact with a biological system (1987) [4-5].

In the current sense, biocompatible biomaterials or materials are living or non-living substances or materials, other than nutrients and medicines, which are brought into contact with living tissues or biological fluids present in the human body and which have the property of being transiently tolerated or permanently by them [6].

A biomaterial is any material, natural or synthetic, that totally or partially replaces a living structure or is a biomedical device that performs or replaces a natural function.

Biomaterials are advanced materials, created to be used as a biocompatible interface with the human body, in the form of medical devices, implants and prosthetic systems [2]. Biomaterials are a class of special materials that are used in contact with biological tissues: blood, cells, proteins and other living substances and that are able to function in close contact with the living organism, with minimal side effects from it [7].

The concept also applies to any material entity that can be used in endogenous (internal) or exogenous (external) contact with the human body for the purpose of treating, modifying, or replacing a tissue, organ, or function. body, but also its diagnosis or monitoring [8].

In the present context, the prefix bio- refers to what is alive or from which life results, especially when an inert object or system is introduced into a living organism.

Biomaterials are used to completely or partially replace the shape and function of diseased tissue or organ, but also to interface with a biological environment, including medical devices for diagnosing and storing drugs or even biochips that could be integrated into computers. Biomaterials assume a significant part not just in the finding and treatment of an infection yet in addition in its ensuing development, on the grounds that upon contact with living matter, the common response of biomaterial $\div$ creature can be helpful or can be destructive [9].

Biomaterial in medical terminology is "any natural or synthetic material (including polymers or metal) and is intended for insertion into living tissues as part of a medical device or implant (artificial or temporomandibular heart, joint replacement, Harrington prosthesis for spine, artificial joint of the hip, device for renal dialysis)". Biomaterials from a health perspective can be defined as "materials that possess new properties that make them suitable for immediate contact with living tissue without causing immune rejection or adverse reaction" [10].

The knowledge gained over the last 70 years has led scientists to model the characteristics of materials at a high level. Thus, tens of thousands of different materials have evolved with specialized characteristics that suit the different service requirements of modern and complex society [2].

The prefix "bio" for biological materials refers to "biocompatible", not "biological" or "biomedical", because it is often misunderstood [3]. The mission of biological materials is to save human lives, ensure medical care and improve the quality of life [4]. Biomaterial's science is part of the broader biomedical engineering discipline. Since engineering and materials science are used to derive the foundations of mathematics, physics, and chemistry, biomedical engineering and biomaterials have also taken biology as their basic science. Thus, the field of parent discipline is expanded in a way that no other activity in the field of engineering currently does or does not rigorously affect biomedical engineering or the biomaterials community. The 21st century is transforming into the era of biology. Biomedical engineering and biomaterials are based on this key scientific advancement to make contributions to society. Biomedical engineering and its biological materials are developing as a discipline worldwide. Mature educational programs such as undergraduate biomedical engineering are constantly emerging around the world, and the Department of Biomedical Engineering is booming [11].

The common feature of biological materials is that they are in close contact with living organisms.

1.2. Characteristics of biomaterials

If it is relatively easy to find a material that meets certain functional requirements, it is very difficult to find one that is able to maintain its performance for a long time without damage, with unwanted effects induced in the body. While bringing a material into the living organic entity, a progression of extremely complex connections can show up, having the option to distinguish four explicit wonders that are unitary in the purported "idea of biocompatibility", specifically [12]:

1) introductory cycles that occur at the biomaterial interface ÷ living tissue and that are firmly identified with the physico - compound cycles that happen in the main minutes of the contact between the biomaterial and the living tissue;

2) the impact that the presence of biomaterial as an unfamiliar body has on the living tissue encompassing the embed, which can be estimated whenever, from a couple of moments to years;

3) the impact that living tissue has on the biomaterial through the progressions saw in the biomaterial, an impact depicted as consumption or debasement;

4) results of the response at the interface that are efficiently seen on the body surface or in specific explicit regions, restoratively perceived as the advancement of explicit hypersensitivities, the inception of growths or the presence of irresistible cycles.

The biomaterial ÷ tissue interface, which is set up by implantation, is definitely a blood ÷ material interface and the underlying occasions are overwhelmed by the ingestion of blood proteins on the embed surface. At this contact it was established that a series of physico-chemical phenomena take place under specific physiological conditions. The precise nature of the mechanism by which foreign surfaces (of biomaterials) initiate blood clotting is not very clear, but in some cases, phenomena leading to this could be highlighted [13].

The presentation of a biomaterial in the living creature decides an embed tissue connection, which can produce clashing responses. They can be poisonous, mechanical, electrochemical and natural. It can even prompt genuine harm deep down or nearby tissue, or to the get together utilized. Because of these marvels, contingent upon the nature of the biomaterial, the spot of implantation and different causes, consumption happens on the outer layer of the embed, with the deficiency of its quality. Contingent upon the clinical application for which it is reasonable, a biomaterial should have at least one of the properties introduced in table 1.1.

Table 1.1. The main characteristics of biomaterials [14].

Characteristic	Comments
Biocompatibility	Biologically compatible with host tissue (eg, must not cause rejection, inflammation and immune responses)
Bioactivity	Easily achieve direct biochemical attachment to host tissue
Biodegradability	The biodegradation time must be adapted to match the time of bone formation
Degradation mode	Surface or depth erosion
Osteoconductivity and osteoinductivity	Ability to support the growth of germline capillaries, mesenchymal perivascular tissues and host osteoprogenitor cells in the three-dimensional structure of the graft that acts as a support
Porous structure	Needed to maximize space for cell adhesion and growth, revascularization, proper nutrition and oxygen supply
Three-dimensional structure	For support in the process of cell growth and in the transport of nutrients and oxygen

The properties of a biomaterial are unequivocal in guaranteeing the biocompatibility of an embed:

- from a substance (compositional) perspective, a biomaterial should not contain components that produce unfavorable and/or incendiary responses upon implantation. A significant angle is additionally identified with the conceivable arrangement on the embed surface, in vivo conditions, of new constructions and pieces, contingent upon the connections that show between the biomaterial and the natural conditions explicit to the implantation region. Their inclination and physico-substance qualities might influence the drawn-out dependability of the embed [15-18].

- according to an underlying perspective, a biomaterial should have a thickness and a porosity relating to the primary capacity that the embed is to satisfy in the organic entity wherein the implantation is made. Of specific significance is the minuscule idea of the embed surface.

- mechanical properties - a biomaterial, contingent upon the capacity that the embed should act in the living life form, should have satisfactory mechanical strength, hardness and unwavering quality.

- on account of visual, dermatological and dental applications, biomaterials should likewise have fitting optical properties.

- one more significant viewpoint is identified with the machinability of the biomaterial, this affecting the designing of the embed itself.

1.3. Classification of biomaterials

Biomaterials are synthetic materials compatible with the human body, with a wide range of properties, which can be transformed into medical devices that correspond to strictly imposed functional parameters.

Biocompatible materials are intended to "work under biological constraint" and thereby become adapted to various medical applications.

In this unique circumstance, a biomaterial is any substance or blend of substances, other than therapeutic items, of manufactured or normal beginning, which can be utilized for an endless period, thusly or as a part of a framework (gadget) for morphological reconstitution furthermore/or practical of tissues, organs [19, 20].

Today, biomaterials mirror the combination of Materials Science, in the entirety of its perspectives, with an exceptionally huge number of organic angles. The field of biomaterials is brimming with studies and conversations about medicines that utilize natural cycles to connect with materials and, thusly, to assist with shaping tissues and recapture physical processes. Indeed, this is the idea of "shrewd materials", materials that invigorate tissue development have turned into a medium in the field as can be seen from the endless investigations including all classes of materials, including metals, ceramics, polymers and composites. The materials have a surface organization intended to communicate with natural parts and cell capacities. Usefulness, according to a natural perspective, is likewise incorporated into material designs or is joined onto material surfaces. New materials are intended to mirror natural designs or capacities, leading to the class of biomimetic materials [5]. The original of biomaterials developed between 1960-1970 and were utilized basically as clinical inserts. The fundamental destinations during the assembling of these biomaterials were to keep a harmony among physical and mechanical properties with insignificant harmfulness to have tissues [21-23].

The standard grouping of engineered biomaterials is performed according to an underlying perspective, as per the classes of materials utilized. There are both biomaterials of normal beginning and blended composite biomaterials acquired by consolidating biomaterials of regular beginning with engineered biomaterials. The principal sorts of engineered biomaterials are metallic, ceramic, polymeric, composite and of natural origin, however they can likewise be separated into a few classifications [24-25]:

- ❖ **Metallic** - Fe, Cr, Co, Ni, Ti, Mo and W, used for implants, are tolerated by living tissues in very small quantities. Medical applications: for intracranial aneurysms, vena cava filters, orthopedic implants.

- ❖ **Polymeric** - materials made by binding smaller molecules by primary covalent bonds in a main chain. Medical applications: implants for the replacement or restoration of human soft tissues - sutures, blood vessels, artificial skin.

- ❖ **Ceramics** - polycrystalline compounds, usually inorganic: metal oxides, carbides, sulfides. Medical applications: dental crowns, dental alveoli, hip joint.

- ❖ **Composites** - are formed from two or more distinct phases with different properties from the homogeneous material. Medical applications: dental composites for fillings, methyl methacrylate, bone cement.

The properties of first-generation biomaterials were:

- adequate mechanical properties;
- corrosion resistance in an aqueous environment;
- lack of toxicity or carcinogenicity in living tissue.

Each group of biomaterials can be classified according to these four categories. Natural materials, such as animal heart valves, are made of proteins and have a repetitive polymer-like structure. Thus, they are considered from the category of polymers [26].

The classification of biomaterials can also be done taking into account the interaction with the host organism:

a. Bioinert materials: do not cause any reaction from the body (e.g., stainless steel);

b. Bioactive materials: interact with surrounding tissues and form bonds (e.g., hydroxyapatite);

c. Biodegradable materials: degrade under the action of a biological agent such as enzymes (e.g., polylactic acid);

d. Bioresorbable / bioabsorbable materials: material residues are removed from cellular activity by phagocytosis (e.g., polyglycolic acid);

e. Bioerodible materials: eroded by physical and chemical processes (e.g., polyortoester).

Second-generation biomaterials have been developed to be also bioactive. Substantially progress has been observed in the application of orthopedic and dental use. Examples include bioactive glasses, ceramics, polymers, ceramic glass and composites. Current developments with biomaterials technology translate into the expansion of a third generation of biomaterials that can stimulate a specific cellular response [27].

The third generation of biomaterials includes several types:

➤ synthetic (metals, polymers, ceramics, composites),

➤ derived from nature (plants, tissues),

➤ semi-synthetic or hybrid.

The choice of the right material is made in accordance with the function to be replaced. Unfortunately, there is no biomaterial that meets all the required requirements.

For example, in the case of applications subjected to force loads (dental or hip implants), the mechanical requirements can only be met by metals. In this case, titanium alloys are the most promising due to their high specific strength and low elastic modulus [4]. Titanium has a strong tendency to passivate and quickly forms a thin layer of titanium oxide in the presence of oxygen. This oxide layer is strongly adherent and stable and is not destroyed under normal physiological conditions. Also, due to this layer, there is no direct contact between titanium and the surrounding tissues and thus prevents the corrosion phenomenon of titanium. However, despite these positive properties, titanium has low bone adhesion compared to other materials such as calcium phosphates whose mechanical properties are unfortunately unsatisfactory for such applications [28-30].

Recent research in the field of biomaterials focuses on the production of biomimetic / bionic materials (materials with properties similar to those in living organisms) and

Advanced Metallic Biomaterials   Materials Research Forum LLC
Materials Research Foundations **118** (2022)   https://doi.org/10.21741/9781644901779

biofunctional materials (complex materials that emulate functional structures of the body, such as muscle fiber).

The human body can be viewed as a construction with different levels: the degree of tissues, the degree of organs or frameworks. According to the perspective of tissues, there are hard tissues and delicate tissues, the last being partitioned into those that come into contact with blood and those that don't come into contact with blood. Concerning human organs, they might benefit, where fitting, from the help gave by biomaterials through inserts or counterfeit organs. Each class of biomaterials enjoys its benefits and drawbacks, their utilization in the execution of various inserts being affected both by the properties of biomaterials and by the useful prerequisites forced on inserts [31].

Among the biotolerant biomaterials, considered to be from the "first generation" of biomaterials, we mention stainless steels and cobalt-based alloys (Co-Cr type). These biomaterials correspond to distant osteogenesis, and a separating layer of connective tissue is formed following the interaction of tissues with metal ions or in other words, a non-adherent fibrous capsule [32].

Bioinert biomaterials (titanium, tantalum, alumina, polyethylene) correspond to contact osteogenesis, which is achieved through intimate contact with a link at the interface between the biomaterial and the host tissue. These biomaterials, from the second generation of biomaterials, have a neutral or inert behavior in the human body, do not have a degenerative action and do not have a significant influence on metabolism [33].

Among the bioinert biomaterials, of unique interest are those with osteotropic structure, which incorporates titanium. These biomaterials, because of the compound and micromorphological biocompatibility with the bone tissue, make a physico-substance association with it, the interface wonder being absorbed with the connection osteogenesis.

Titanium amalgams are being utilized to an ever-increasing extent, because of the need to supplant treated steels and cobalt-based combinations that have impediments being used, produced by certain inadequacies of biocompatibility with human tissues. These lacks are brought about by certain components present in their synthetic piece (e.g., nickel), which have a harmful activity on human tissues, causing fiery hypersensitive responses or embed dismissal responses [34].

For bioactive biomaterials (calcium phosphate, ceramic bottles, hydroxyapatite) is considered typical binding osteogenesis, based on the appearance of a chemical bond between biomaterial and tissue. Considered to be of the third generation of biomaterials, they are designed to be introduced into the metabolic process and to stimulate tissue growth.

Ceramic bottles and ceramics containing oxides of Si, Na, Ca and P (SiO_2, NaO_2, CaO and P_2O_5) seem to be the only biomaterials known to form a chemical bond with bone tissue, following a strong, mechanical bond, tissue bone-implant. These biomaterials are called bioactive due to their connection with hard tissues (bone tissues), but in some cases also with soft tissues, being time dependent and introducing surface kinetic changes by implanting them in living tissues. In particular, the ion exchange reaction between an implant made of a bioactive material and the surrounding human fluids results in the formation of a biologically active film on the implant surface, which is chemically and crystallographically equivalent to the mineral phase. of bone tissue. This equivalence underlies the relatively strong interphase connection between the implant and it. Although bioactive materials seem to be the ideal answer to implant fixation problems, they are not available for applications that require high loads or good shock resistance [35-37].

The mechanism of the formation of a new bone surface following the interaction of bone tissue with a bioactive ceramic biomaterial is particularly interesting. Immediately after the implantation of a bioactive ceramic implant, an ion exchange takes place between it and the human environment, the diffusion of ions having a double meaning. After a while, the result is the formation of a "new bone" film.

Bioresorbable biomaterials, such as tricalcium phosphate or the polylactic acid-polyglycolic acid copolymer, are used for the temporary replacement of tissues and are intended to be replaced slowly, over time, by the recovering tissues. They are also used in pharmaceutical applications.

These biomaterials, considered to be part of the fourth generation of biomaterials, are currently being studied extensively. They are often used as a carrier system for cell proliferation and differentiation, and are intended to support current human organ transplants. Also, applications related to tissue engineering have the role of supporting the functions of human tissues using appropriate biomaterials or tissue formations obtained from cell cultures obtained in vitro [38].

1.3.1. Metallic materials

Metals have been used, in various forms, in implants. The first metal used especially for use in implants in the human body was vanadium ("Sherman Vanadium Steel"), which was used in the manufacture of plates and screws used in bone fractures. Most metals used to make implants, such as Fe, Cr, Co, Ni, Ti, Ta, Mo and W, can be tolerated by the body only in small amounts. Sometimes these metallic elements, in their natural form, are essential in cellular functions (Fe) or in the synthesis of vitamin B12 (Co), but they cannot be tolerated in large quantities in the body. The biocompatibility of metal implants is a considerable problem because they tend to corrode in a hostile environment. The consequence of corrosion is the loss of material, which will weaken the resistance of the implant, and perhaps even more so, corrosion causes tissue damage, leading to unwanted effects. Next, the relationship between the composition, structure and properties of metals and alloys used to manufacture implants is studied [39].

The use of metal materials for medical implants can be traced back to the 19th century, when the era of the metal industry began to expand during the Industrial Revolution. Metallic materials are generally grouped as 'ferrous and non-ferrous'.

Ferrous materials represent more than 50% of the class of metallic materials. Depending on their importance and subsequent applicability, ferrous materials are divided into several subcategories: iron and alloy steel, stainless steel, cast iron. In the 'non-ferrous materials' family, the subdivision is again based on importance and use. In this subdivision there are several small groups, such as materials with high melting temperature, super alloys, shape memory alloys, glass materials, soldering materials, noble metals, etc. The development of metal implants was determined primarily by requests to address bone repair, usually internal fracture, fixation of long bones. In any case, practically no endeavors to embed metal gadgets, for example, spine wires and bone substitution bars made of iron, gold or silver, were effective until Lister's careful aseptic, the procedure being executed in 1860s. From that point forward, metallic materials have prevailed in muscular medical procedure, assuming a significant part in most muscular gadgets, including impermanent gadgets (bone plates, pins and screws) and super durable inserts (absolute joint substitutions). Simultaneously, metals with applications in dental and orthodontic practice have been discovered, including teeth, fillings and roots [40].

As of late, metallic biomaterials have been presented in the utilization of unusual reconstructive medical procedure of hard tissues/organs, for example, the utilization of NiTi shape memory composites as vascular stents and the improvement of new magnesium-based compounds for bones, tissue designing and recovery. Regardless of the enormous number of metals and amalgams equipped for being delivered in industry, a couple are biocompatible and have long haul accomplishment as an embed material. Biomaterials are components in most commercially available orthopedic medical devices. They can be classified into the following four groups based on the main alloying element:

- stainless steels,
- cobalt-based alloys,
- titanium-based alloys,
- other alloys (alloys based on Mg, Fe and Ta) [41-43].

1.3.2   Polymeric materials

Polymers are classified according to several criteria, as follows [44-46]:

➤ from the point of view of the polymer composition are known:

- homopolymers (composed of the same types of structural units): polyoxymethylene, natural rubber, polyethylene, polystyrene, etc.

- copolymers (composed of different monomers): proteins, nucleic acids, butadiene-styrene rubber, etc.

➤ from the point of view of the arrangement of atoms or groups of atoms in macromolecular compounds, they are classified in:

- linear or filiform polymers in the form of a long macromolecular chain, which grew in the only direction (e.g., cellulose)

- branched polymers (e.g., amylopectin)

- three-dimensional polymers (e.g., vulcanized rubber)

➤ according to their origin they are classified in:

- natural polymers (natural rubber, cellulose, starch, protein substances, nucleic acids, etc.).

- artificial polymers (such as rubber, nitrocellulose, triacetylcellulose).

- synthetic polymers (polyethylene, polystyrene, butadiene rubber, etc.)

➤ according to the way of use the polymers are classified in:

- elastomers (natural rubber, synthetic rubber);

- plastics (polyethylene, polystyrene); fibers (capron, cotton, linen, polyester, etc.);

- resins (phenolformaldehyde, kinetic glues, etc.).

➤ after the behavior towards the temperature of the polymer it is classified in:

- thermoplastic polymers;

- thermoreactive polymers.

Polymeric materials have been utilized since ancient occasions. Normal polymers are plentiful in nature, found in every living framework and materials like wood, paper, calfskin, regular filaments. While normal polymers hold their characteristic person, engineered materials are generally utilized. The first artificial polymers, obtained by chemical modification of natural substances, were produced in the second half of the nineteenth century [47].

Synthetic polymers (plastics of modern society) were developed in the twentieth century, mostly during the years 1950-1970, determined by the expansion of the chemical industry. The vast majority of plastics are derived from oil, and environmental concerns have led to more recent evolutions of polymers from renewable resources. The chemical entity is known as the repeat unit and defines the structure and properties of the polymer. The repeated unit usually consists of small molecules called monomers [48].

The monomers must have one or more functional groups or unsaturated chemical bonds to produce a polymer. These polymers that have the same unit repeated throughout the molecule are known as homopolymers and are the most common polymeric materials. Medical equipment, medical devices and drug delivery systems have seen a significant use and development of polymeric materials, especially bio-absorbable systems. The primary benefits of polymers incorporate modifiable mechanical properties just as debasement energy. Because of the great crystallinity, poly-α-hydroxy acids have high rigidity and solidness in natural solvents. Certain poly-α-hydroxy acids likewise have astounding fiber-shaping limit and, accordingly, have been utilized in the advancement of stitch materials. The fundamental detriment is that the items coming about because of the corruption decide the decrease of the neighborhood pH. This isn't simply conceivable to speed up the degradation process, yet in addition to actuate a fiery response [49-50].

1.3.3. Ceramic materials

Ceramics are compounds of metal elements ionically and / or covalently bonded to non-metallic elements. They have a wide range of uses, including filling bone defects, fixing fractures and prosthetic coatings.

The characteristics of ceramic materials include hardness, high mechanical strength, high compressive strength, high rigidity, corrosion resistance and wear. They usually work well at compressive forces, but behave poorly at tensile forces. Ceramic materials are sensitive to plastic deformation leading to cracks and defects.

Over the years ceramic materials have developed and include a wide range of materials that can be divided into three main categories:

- bioinert (alumina, zirconium),
- bioactive (glass ceramics),
- bioresorbable (calcium phosphates) [51].

Calcium phosphate biomaterials are accessible in different actual structures including hydroxyapatite and tricalcium phosphate. Hydroxyapatite, in spite of the fact that it is osteoconductive (causes development bone on a surface) and osteoinductive (advancing osteogenesis) has an exceptionally low biodegradability and is in this way not a practical material according to a biodegradable perspective.

Tricalcium phosphate has the same characteristics as hydroxyapatite, but also has a much higher rate of degradation compared to it. A ceramic material is defined as a non-metallic inorganic solid material, but in a narrow sense, it is a solid solid material obtained by the process of heat treatment of inorganic powders. Ceramic materials have been used since antiquity. Pottery has a long history. One of the first ceramic materials used by humans was a stone tool. Obsidian in composition, such a material can be classified as glassy. Similarly, porcelain and earthenware also have a long history. The same can be said of glass; human beings have been producing solid glass since the years 3000 BC. Ceramic powder was also used as an adhesive in the construction of the pyramids. Ceramics are classified, from the point of view of structure, as follows: unique crystal, sintered materials, glass, powders [52-54].

Ceramic materials can be handled by blending particles of material in with water and a natural cover. The microstructure of the end result relies upon the hotness cycle, the most extreme temperature came to and the span of the hotness treatment applied, known as sintering. The sintering system is vital in the creation of biodegradable earthenware production to guarantee that the materials got the properties vital for biomedical use. The primary benefit of biodegradable pottery is the permeable design and the speed of biodegradation. Because of the permeable construction, bone development can happen on the outer layer of the ceramic because of the interconnection of pores that works with tissue ingestion giving a pathway to wholesome vehicle. The presence of pores is subsequently significant for the osteoconductive trademark. The fundamental hindrance of biodegradable pottery is their delicacy and frail elasticity, dropping their utilization as burden bearing inserts [55, 56].

1.3.4. Composite materials

Mechanical contest between countries is to make a superior way of life to benefit humanity. Materials assume a significant part in propelling human existence. Materials utilized in item advancement rely upon accessibility, execution and cost. Composite materials have supplanted numerous customary materials: metals, pottery, wood and polymers because of the benefits they offer: mechanical strength and inflexibility erosion obstruction, breaking strength, wear opposition, high temperature obstruction, surface hardness, dimensional security, limit vibration damping, somewhat low cost and so forth Lately, there is a fast development in the turn of events and utilization of composite materials because of the benefits it offers, like low thickness and erosion opposition. Composites contain at least two stages, in which the constituent materials are typically handled independently and

15

afterward joined. The properties of composites are not the same as the properties of constituent materials. One of the constituents is the built-up stage as filaments, particles or sheets and are fused into the other constituent, called the lattice [57,58]. Composites are strong materials made out of a grid that encompasses and ties the sinewy fortifications. Composite materials are viewed as the materials of today, as they are presently applied in different businesses from youngsters' toys to airplane bodies. They can be viewed as the materials of tomorrow because of the materialness they can present later on, for example, in the field of nanocomposites, utilitarian materials and astute materials. The most generally utilized lattice is polymer or plastics (polymer grid composite, PMC), less significantly, pottery (clay network composite, CMC) and metals (metal framework composite, MMC). Composites are anisotropic materials, having various properties along the tomahawks in various ways. A wide scope of networks and fittings can be joined in different structures with for all intents and purposes limitless changes [59].

References

[1] Ratner, B.D.; Hoffman, A.S.; Schoen, F.J.; Lemons, J.E. (Eds.) Biomaterials Science: An Introduction to Materials in Medicine, 2nd ed.. *Academic Press: San Diego*, CA, USA, **2004**.

[2] Dumitraşcu, N. Biomateriale şi biocompatibilitate. *Ed. Universităţii "Alexandru Ioan Cuza"*, Iaşi, **2007**.

[3] Bronzino, J.D.; Peterson, D.R. (Eds.) The Biomedical Engineering Handbook, 4th ed.; *CRC Press: Boca Raton*, FL, USA, **2015**.

[4] Williams, D.F. (Ed.) Definitions in Biomaterials—Proceedings of a Consensus Conference of the European Society Biomaterials. *Elsevier: New York*, NY, USA, 1987; Volume 4.

[5] Williams, D.F.; Black, J.; Doherty, P.J. Consensus report of second conference on definitions in biomaterials. In Biomaterial-Tissue Interfaces; Doherty, P.J., Williams, R.L., Williams, D.F., Lee, A.J.C., Eds.; *Elsevier: Amsterdam, The Netherlands* **1992**, *10* , 525–533.

[6] Thibodeau, G.A.; Patton, K.T. Anatomy and Physiology, 4th ed.; *Mosby: MO*, USA **1999**, 851–887. ISBN 0-323-00192-0.

[7] Blackwood, D.J. Biomaterials: Past successes and future problems. *Corros. Rev.* **2003**, *21*, 97–124. https://doi.org/10.1515/CORRREV.2003.21.2-3.97

[8] Virtanen, S. Corrosion of biomedical implant materials. *Corros. Rev.* **2008**, *26*, 147–171. https://doi.org/10.1515/corrrev.2008.147

[9] Witte, F.; Eliezer, A. Biodegradable metals. In Degradation of Implant Materials; Eliaz, N., Ed.; *Springer: New York*, NY, USA, **2012**, *Chapter 5*, 93–109. https://doi.org/10.1007/978-1-4614-3942-4_5

[10] Laing, P.G. Compatibility of biomaterials. Orthop. Clin. North Am. **1973**, *4*, 249–273. https://doi.org/10.1016/S0030-5898(20)30792-6

[11] Thomsen, P.; Ericson, L.E. Inflammatory cell response to bone implant surfaces. In The Bone-Biomaterial Interface; Davis, J.E., Ed.; *University of Toronto Press: Toronto*, ON, Canada, **1991**, 153–164. https://doi.org/10.3138/9781442671508-018

[12] Eliaz, N. Degradation of Implant Materials; *Springer: New York*, NY, USA, **2012**. https://doi.org/10.1007/978-1-4614-3942-4

[13] Eliaz, N. Biomaterials and corrosion. In Corrosion Science and Technology: Mechanism, Mitigation and Monitoring; Kamachi Mudali, U., Raj, B., Eds.; *Narosa Publishing House: New Delhi*, India, **2008**, *Chapter 12*, 356–397.

[14] Black, J. Biological Performance of Materials—Fundamentals of Biocompatibility, 2nd ed.; *Marcel Decker: New York*, NY, USA, **1992**, 38–59.

[15] Winter, G.D. Tissue reactions to metallic wear and corrosion products in human patients. *J. Biomed. Mater. Res. Symp.* **1974**, *5*, 11–26. https://doi.org/10.1002/jbm.820080304

[16] Mears, D.C. Materials and Orthopaedic Surgery; *Williams & Wilkins: Baltimore*, MD, USA, **1979**, *Chapter 1*.

[17] Lundström, I.M.C. Allergy and corrosion of dental materials in patients with oral lichen planus. *Int. J. Oral Surg.* **1982**, *12*, 1. https://doi.org/10.1016/S0300-9785(83)80060-X

[18] Banoczy, J.; Roed-Petersen, B.; Pindborg, J.J.; Inovay, J. Clinical and histologic studies on electrogalvanically induced oral white lesions. Oral Surg. *Oral Med. Oral Pathol.* **1979**, *48*, 319–323. https://doi.org/10.1016/0030-4220(79)90031-8

[19] Eliaz, N. Wear particle analysis. In *ASM Handbook*: Friction, Lubrication, and Wear Technology; Totten, G.E., Ed.; ASM International: Materials Park, OH, USA, **2017**, *18*, 1010–1031. https://doi.org/10.31399/asm.hb.v18.a0006383

[20] Eliaz, N.; Hakshur, K. Fundamentals of tribology and the use of ferrography and bio-ferrography for monitoring the degradation of natural and artificial joints. In Degradation of Implant Materials; Eliaz, N., Ed.; **Springer: New York, NY, USA**, **2012**, *Chapter 10*; 253–302. https://doi.org/10.1007/978-1-4614-3942-4_10

[21] Yang, L.; Webster, T.J. Biological response to and toxicity of nanoscale implant materials. In Degradation of Implant Materials; Eliaz, N., Ed.; *Springer: New York*, NY, USA, **2012**, *Chapter 18*, 481–508. https://doi.org/10.1007/978-1-4614-3942-4_18

[22] Venugopalan, R.; Gaydon, J. A Review of Corrosion Behaviour of Surgical Implant Alloys; Technical Review Note 99–01; *Perkin Elmer Instruments: Princenton*, NJ, USA, **2001**, 99–107.

[23] Implants for Surgery—Metallic Materials—Part 1: Wrought Stainless Steel; ISO 5832-1; International Organization for Standardization: Geneva, Switzerland, **2016**.

[24] Bhat, S.V. Biomaterials; **Narosa Publishing House: New Delhi**, India, **2002**, 36–38.

[25] Hastings, G.; Black, J.; Murphy, W. (Eds.) Handbook of Biomaterial Properties, 2nd ed.; *Springer Science+Business Media: New York*, NY, USA, **2016**.

[26] Park, B.J.; Park, J.-C. Biological safety evaluation of polymers. In Degradation of Implant Materials; Eliaz, N., Ed.; *Springer: New York*, NY, USA, **2012**, *Chapter 17*, 463–479. https://doi.org/10.1007/978-1-4614-3942-4_17

[27] Kuhn, A.T.; Neufeld, P.; Rae, T. Synthetic environments for testing of metallic biomaterials. In The Use of Synthetic Environments for Corrosion Testing, ASTM STP 970; Francis, P.E., Lee, T.S., Eds.; *ASTM: Philadelphia*, PA, USA, **1988**, 79–95. https://doi.org/10.1520/STP26001S

[28] Hallab, N.J.; Jacobs, J.J. Orthopedic implant fretting corrosion. *Corros. Rev.* **2003**, *21*, 183–214. https://doi.org/10.1515/CORRREV.2003.21.2-3.183

[29] Anderson, J.M.; Rodriguez, A.; Chang, D.T. Foreign body reaction to biomaterials. *Semin. Immunol.* **2008**, *20*, 86–100. https://doi.org/10.1016/j.smim.2007.11.004

[30] Solar, R.J. Corrosion resistance of titanium surgical implant alloys, a review. Corrosion and degradation of implant materials. In Corrosion and Degradation of Implant Materials, ASTM STP 684; Syrett, B.C., Acharya, A., Eds.; *ASTM: Baltimore*, MD, USA, **1979**, 259–273. https://doi.org/10.1520/STP35949S

[31] Manivasagam, G.; Dhinasekaran, D.; Rajamanickam, A. Biomedical implants: Corrosion and its prevention—A review. *Recent Patents Corros. Sci.* **2010**, *2*, 40–54. https://doi.org/10.2174/1877610801002010040

[32]   Pound, B.G. Corrosion behavior of metallic materials in biomedical applications. I. Ti and its alloys. *Corros. Rev.* **2014**, *32*, 1–20. https://doi.org/10.1515/corrrev-2014-0007

[33]   Gilbert, J.L.; Mali, S. Medical implant corrosion: Electrochemistry at metallic biomaterial surfaces. In Degradation of Implant Materials; Eliaz, N., Ed.; *Springer: New York*, NY, USA, **2012**, *Chapter 1*, 1–28. https://doi.org/10.1007/978-1-4614-3942-4_1

[34]   Virtanen, S. Degradation of titanium and its alloys. In Degradation of Implant Materials; Eliaz, N., Ed.; *Springer: New York*, NY, USA, **2012**, *Chapter 2*, 29–55. https://doi.org/10.1007/978-1-4614-3942-4_2

[35]   Baltatu, M.S.; Tugui, C.A.; Perju, M.C.; Benchea, M.; Spataru, M.C.; Sandu, A.V.; Vizureanu P. Biocompatible Titanium Alloys used in Medical Applications. *Revista de Chimie* **2019**, *70(4)*, 1302-1306. https://doi.org/10.37358/RC.19.4.7114

[36]   Bălțatu, M.S.; Vizureanu, P.; Mareci, D.; Burtan, L.C.; Chiruță, C.; Trincă, L.C. Effect of Ta on the electrochemical behavior of new TiMoZrTa alloys in artificial physiological solution simulating in vitro inflammatory conditions. *Materials and Corrosion* **2016**, *67(12)*, 1314-1320. https://doi.org/10.1002/maco.201609041

[37]   Bălțatu, M.S.; Vizureanu, P.; Istrate, B. Physical and structural characterization of Ti-based alloy, *International Journal of Modern Manufacturing Technologies* **2015**, *VII (2)*, 12-17.

[38]   Chelariu, R.; Bujoreanu, G.; Roman, C. Materiale metalice biocompatibile cu baza titan, *Ed. Politehnium*, Iași, **2006**.

[39]   Zierold, A.A. Reaction of bone to various metals. *Arch. Surg.* **1924**, *9*,   365–412. https://doi.org/10.1001/archsurg.1924.01120080133008

[40]   Eliaz, N.; Kamachi Mudali, U. (Eds.) Special Issue: Biomaterials Corrosion. *Corros. Rev.* **2003**, *2(2-3)*.

[41]   Tengvall, P.; Lundstrom, I. Physico-chemical considerations of titanium as a biomaterial. *Clin. Mater.* **1992**, *9*, 115–134. https://doi.org/10.1016/0267-6605(92)90056-Y

[42]   Hanawa, T. Degradation of dental implants. In Degradation of Implant Materials; Eliaz, N., Ed.; *Springer: New York*, NY, USA, **2012**, *Chapter 3*, 57–78. https://doi.org/10.1007/978-1-4614-3942-4_3

[43]   Ciolac, S.; Vasilescu, E.; Drob, P.; Popa, M.V.; Anghel, M. Long-term in vitro study of titanium and some titanium alloys used in surgical implants. *Rev. Chim.* **2000**, *51*, 36–41.

[44]   Bunea, D.; Nocivin, A. Materiale biocompatibile, *Ed. şi Atelierele Tipografice Bren*, Bucureşti, **1998**.

[45]   Popa, C.; Cândea, V.; Şimon, V.; Lucaciu, D.; Rotaru O., Ştiinţa biomaterialelor, *Ed. U.T. Press*, Cluj-Napoca, **2008**.

[46]   Chen, Q. Thouas G.A., Metallic implant biomaterials, *Materials Science and Engineering R*, **2015**, *87*, 1–57. https://doi.org/10.1016/j.mser.2014.10.001

[47]   Baier, R.E. Surface behavior of biomaterials: the theta surface for biocompatibility, *Journal of material science: Materials in medicine* **2006**, *17(11)*, 1057-1062. https://doi.org/10.1007/s10856-006-0444-8

[48]   Miculescu, F. Tehnici de analiză şi control a biomaterialelor, Ed. Printech, Bucureşti, **2009**.

[49]   Roşu, R.A. Metode de obţinere şi de prelucrarea biomaterialelor pentru proteze umane-teză de doctorat, Universitatea "Politehnica", Timişoara, **2008**.

[50]   Bombac D.M., Brojan M., Fajfar P., Kosel F., Turk R., Review of materials in medical applications, *Materials and Geoenvironment*, **2007**, *54(4)*, 471-499.

[51]   Eliaz, N.; Metoki, N. Calcium phosphate bioceramics: A review of their history, structure, properties, coating technologies and biomedical applications. *Materials* **2017**, *10*, 334. https://doi.org/10.3390/ma10040334

[52]   Guslitzer-Okner, R.; Mandler, D. Electrochemical coating of medical implants. In Applications of Electrochemistry in Biology and Medicine I. *Modern Aspects of Electrochemistry* **2011**, *52*, 291 - 342. https://doi.org/10.1007/978-1-4614-0347-0_4

[53]   Eliaz, N.; Sridhar, T.M. Electrocrystallization of hydroxyapatite and its dependence on solution conditions. *Cryst. Gr. Des.* **2008**, *8*, 3965–3977. https://doi.org/10.1021/cg800016h

[54]   Eliaz, N.; Sridhar, T.M.; Kamachi Mudali, U.; Raj, B. Electrochemical and electrophoretic deposition of hydroxyapatite for orthopaedic applications. *Surf. Eng.* **2005**, *21*, 238–242. https://doi.org/10.1179/174329405X50091

[55]   Kokubo, T.; Yamaguchi, S. Bioactive metals prepared by surface modification: Preparation and properties. In Applications of Electrochemistry in Biology and

Medicine I; (Modern Aspects of Electrochemistry, No. 52); Eliaz, N., Ed.; *Springer Science+Business Media: New York*, NY, USA, **2011**, 377–421.

[56]   Geuli, O.; Metoki, N.; Eliaz, N.; Mandler, D. Electrochemically driven hydroxyapatite nanoparticles coating of medical implants. Adv. Funct. Mater. **2016**, *26*, 8003–8010. https://doi.org/10.1002/adfm.201603575

[57]   Thomas, M.B.; Metoki, N.; Geuli, O.; Sharabani-Yosef, O.; Zada, T.; Reches, M.; Mandler, D.; Eliaz, N. Quickly manufactured, drug eluting, calcium phosphate composite coating. *Chem. Sel.* **2017**, *2*, 753–758. https://doi.org/10.1002/slct.201601954

[58]   Metoki, N.; Il-Baik, S.; Isheim, D.; Mandler, D.; Seidman, D.N.; Eliaz, N. Atomically resolved calcium phosphate coating on a gold substrate. Nanoscale **2018**, *10*, 8451–8458. https://doi.org/10.1039/C8NR00372F

[59]   Antoniac I. Biomateriale metalice utilizate la executia componentelor endoprotezelor totale de sold, *Ed. Printech*, Bucureşti, **2007**.

Advanced Metallic Biomaterials
Materials Research Foundations **118** (2022)

Materials Research Forum LLC
https://doi.org/10.21741/9781644901779

CHAPTER 2

Titanium alloys

Titanium is currently one of the most widely used metals as a basis for obtaining metallic materials with special properties. Polymorphism, good alloying ability with many elements of Mendeleev's system, the formation of a wide range of solid solutes or intermetallic phases with varying solubility, high melting temperatures and excellent corrosion resistance create favorable conditions for a wide variety of structures and properties [1,2].

It was found in 1788 in the form of dioxide and was extracted in the form of a powder in 1825 contaminated with nitride, and in order to obtain a metal of suitable purity it took 100 years until 1925.

The literature began to pay excessive attention to titanium and its alloys only in 1940, when the results obtained in the processing of malleable titanium, extracted by the decomposition of iodide, are communicated. Subsequently, the method of magnesium thermal reduction of tetrachloride in a neutral atmosphere or in vacuum was spread on an industrial scale, making the titanium sponge which alloyed with other metals and elaborated in vacuum electric furnaces with consumable electrodes gave the possibility to make alloys of this metal [3,4].

Titanium is one of the main components of the earth's crust, in a percentage of 30%. Due to the difficulties of procuring it from ores, it was considered a rare metal.

The color of thiatnum is silver white, it has breaking strength between 500 and 880 MPa, it has melting temperatures of $1668°C$ and boiling temperature of $3027°C$, the weight is specific to titanium, the elongation is between 4 and 28% and the hardness of 180-260 HB.

Titanium is part of the fourth subgroup of Mendeleev's system along with thorium, hafnium and zirconium, with order number 22 and an atomic weight of 47.90. Although it is one of the most widespread elements in the earth's crust, it has long been little studied, considering that it is a hard and brittle metal and can only be used in technology as a deoxidizer [5].

On heating, titanium expands 2.5 times less than aluminum, and its electrical resistivity is 5 times higher than that of iron and 20 times greater than that of aluminum.

Titanium was discovered in 1791 by amateur and pastoral geologist William Gregor. of TiO_2 and the rest of FeO, which is found in metamorphic and magmatic rocks and crystallizes in the ramboendric system. it was behind a magnet. The analysis of the sand determined the presence of two metal oxides: iron oxide (which explains the attraction to the magnet) and 42.25% a white metal oxide that could not be identified [6].

Gregor, realizing that the unknown oxide contained a metal that did not match the properties of any other element known at the time, reported to the Royal Geological Society of Cornwall and Crell's Annalen Journal of Science [7].

During constant amount Franz-Joseph Muller von Reichenstein made an identical substance, however couldn't determine it. The compound was then rediscovered in 1795 by the German chemist Martin Heinrich Klaproth in his routine in Hungary. Martin Heinrich Klaproth recognized a brand-new component there and named it once the Titans in Greek mythology. once hearing concerning Gregor's earlier discovery, he obtained a sample of manaccanite and confirmed that the ascetic contained atomic number 22.

Pure metal atomic number 22 (99.9%) was initial ready in 1910 by Matthew Hunter by heating TiCl4 with atomic number 11 in a very steel capsule to 700-800°C by Orion method. The metal wasn't used outside the laboratory till 1932, once William Justin Kroll established that it will be made by reducing atomic number 22 chemical compound within the presence of Ca. Eight years later, he formed the method, victimization atomic number 12 or maybe atomic number 11 in what became referred to as the Kroll method. though analysis into a lot of economical and cheaper processes continues (e.g., FFF Cambridge), the Kroll method continues to be used for industrial production (Figure 2.1) [6-8].

Figure 2.1. A titanium crystal bar produced by the iodide process [8].

High purity titanium become synthetic in small portions whilst Anton Eduard van Arkel and Jan Hendrik de Boer determined the method of iodine or crystal bar in 1925 through reacting with iodine and breaking down vapors fashioned over a warm filament into natural metallic.

In the Nineteen Fifties and 1960s, the Soviet Union pioneered the usage of titanium in navy and submarine applications (Alpha elegance and Mike elegance) as a part of Cold War-associated programs.

Beginning withinside the early Nineteen Fifties, the metallic debuted for navy aviation purposes, in particular excessive-overall performance jet plane, beginning with plane inclusive of the F100 Super Saber and Lockheed A-12.

In the United States, the Department of Defense become made aware about the strategic significance of the metallic and supported the primary efforts to marketplace it. During the Cold War, titanium become taken into consideration a strategic cloth through the United States Government, and a big garage of sponges Titanium become maintained through the National Defense Stockpile Center, which become in the end depleted in 2005. Today, the world's biggest manufacturer, VSMPO-Avisma primarily based totally in Russia is predicted to account for 29% of globalwide marketplace share [9].

In 2006, the US Defense Agency awarded $5.7 million to a -consortium marketing campaign to increase a brand-new titanium-making method withinside the shape of metallic powder. Under warmness and stress conditions, the powder may be used to create robust and mild objects, starting from armor to additives for the aerospace, transportation and chemical processing industries.

Titanium is a chemical detail with the image Ti and atomic wide variety 22. It has a low density and is a difficult, bright and corrosion-resistant transition metallic with a silver color. Titanium may be utilized in aggregate with iron, vanadium, molybdenum a good way to produce robust and mild alloys for aerospace (jet engine, projectile or spacecraft), navy use, commercial processes (chemical and petrochemical, paper), automobiles, agri-food, clinical prostheses, orthopedic implants and dental files, dental implants, jewelry, cell telephones and different applications) [10].

Its maximum not unusualplace compound, titanium dioxide, is used withinside the manufacture of white pigments. Other compounds encompass titanium tetrachloride ($TiCl_4$, utilized in writing withinside the sky and as a catalyst) and titanium trichloride ($TiCl_3$, used as a catalyst withinside the production method of apolypropylene).

Two of the maximum beneficial homes of the metallic is the corrosion resistance and the very best hardness-to-weight ratio of all metals. In its natural state, titanium is as difficult as a few sorts of metallic however forty-five instances lighter. There are allotropic paperwork and 5 herbal isotopes of this detail: from ^{46}Ti to ^{50}Ti, with ^{48}Ti being the maximum considerable (73.8%). The homes of titanium are comparable chemically and bodily to the ones of zirconium.

Titanium is constantly associated with different factors in nature (Figura 2.2). It is the 9th maximum considerable detail withinside the earth's crust (0.63% after meals) and the 7th of the metals. It is found in maximum volcanic and sedimentary rocks derived from them, in addition to in creatures or herbal accumulations of water. In fact, of the 801 sorts of volcanic rocks analyzed through the US Geological Survey, 784 contained titanium. The share wherein it's miles determined in soils is about 0.5% to 1.5%. Titanium is extensively disbursed and is determined clearly in particular withinside the minerals anatase, brookite, ilmenite, perovskite, rutile, titanite, however additionally in lots of iron ores. Of these, simplest rutile and ilmenite are of financial significance, even though locating them in excessive concentrations is difficult. Considerable portions of titanium emperors are determined in Western Australia, Canada, China, India, New Zealand, Norway and Ukraine and 4 three million lots of titanium dioxide. Total titanium reserves had been predicted to exceed six hundred million tonnes (Table 2.1) [11].

Meteorites can also additionally comprise this detail, which has been detected withinside the solar and in M-kind stars, the coldest form of star, with a floor temperature of 3200°C (5792° F). Rocks added returned from the moon in the course of the Apollo 17 task are composed of 12.1% TiO_2. Titanium also can be determined in coal ash, vegetation or maybe the human body [12].

Table 2.1. The abundance of titanium on the planet [9,10].

Producer	Thousands of tons	% of total
Australia	1291.0	30.6
South Africa	850.0	20.1
Canada	767.0	18.2
Norway	382.9	9.1
Ukraine	357.0	8.5
Other countries	573.1	13.6
Total planet	4221.0	100.0

Titanium has two allotropic states: Tiα stable up to 882°C and Tiβ stable from this temperature to melting temperature. Titanium has a transformation similar to the martensitic one, being heat treated due to allotropic transformations. By hardening at temperatures higher than the allotropic transformation temperature, a fine needle-like structure similar to martensite is obtained, and after a high return to the temperature immediately below the allotropic point, the structure globalizes (Figure 2.2) [13].

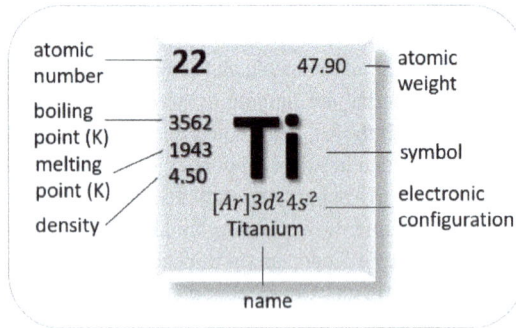

Figure 2.2. The chemical element titanium [13].

Attempts to use titanium in the manufacture of implants date back to the 1930s. Titanium was found to be tolerated in the cat's femur, as were stainless steels and Vittallium (CoCrMo). The fact that it is a light material ($4.5g/cm^3$ compared to $7.9g/cm^3$ for type 316, $8.3g/cm^3$ for cast CoCrMo and $9.2g/cm^3$ for forged CoNiCrMo alloy) and the mechanical-chemical properties good are ideal characteristics for making implants (Figure 2.3).

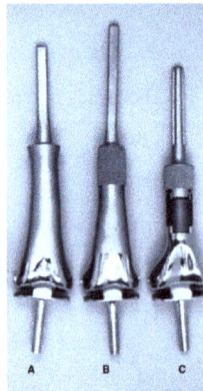

Figure 2.3. (A) A custom endoprosthesis featuring a body casted by the lost-wax method and used from 1980 to 1985 is shown. The stem, based on the Zickel nail design, was welded to the body. (B) An extramedullary porous coating was added to the prosthesis used from 1985 to 1990, which acts as a scaffolding for soft tissue ingrowth. This presumably protects the stem from debris wear and may reduce aseptic loosening. (C) A contemporary modular endoprosthesis used since 1990 features titanium segments and Co-Cr-Mo alloy Morse tapers that prevent cold welding [14].

2.1. Properties

Titanium is a hard, low-density metal that is quite ductile, glossy and silvery white in color. The relatively high melting point temperature (above 1649°C) makes it useful as a refractory material. Commercial types of titanium (with a purity of 99.2%) have a maximum tensile strength of 434 MPa, identical to that of low-quality steel alloys, but are 45% lighter. Titanium is 60% denser than aluminum, but more than twice as strong as the most commonly used 6061-T6 aluminum alloy. These titanium alloys (e.g., Beta C) have a tensile strength of over 1400 MPa. However, the metal loses its hardness when heated to 430°C (806°F).

From the point of view of allotrope and twin crystals, metallic titanium with an alpha hexagonal shape becomes a β-centered cube at 882°C (1619.6°F). As long as it is heated to this trading temperature, the specific heat of the α form will increase significantly, but then it will decrease and stay the same regardless of temperature.

The chemical properties of titanium are its excellent corrosion resistance; it is almost as resistant as platinum, and can withstand the erosion caused by acid or chlorine dissolved in water, but it is soluble in concentrated acids.

Despite the fact that the Pourbaix diagram for titanium shows that it is, from a thermodynamic point of view, a highly reactive metal, its reactions with water and air are slow (Figure 2.4) [15].

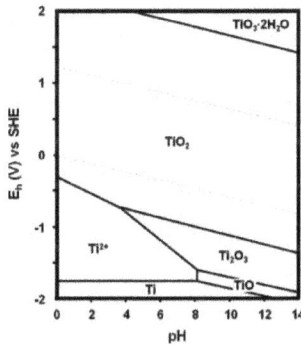

Figure 2.4. Pourbaix diagram for titanium in water, perchloric acid or sodium hydroxide [16].

Titanium is heated in air to 1200°C (2192°F) and burned in pure oxygen at 610°C (1130°F) or higher to form titanium dioxide. Therefore, titanium cannot be melted in the open air, because it burns before it reaches the melting point, so this process can only be carried out in an inert atmosphere or in a vacuum. Titanium is resistant to dilute sulfuric acid and

Materials Research Forum LLC
https://doi.org/10.21741/9781644901779

hydrochloric acid, chlorine gas, chlorine solution and most organic acids. It is paramagnetic (weakly attracted by magnets) and has relatively low electrical and thermal conductivity (Table 2.2) [17].

Table 2.2. Physical properties of titanium [11].

Characteristic	Value
Atomic number	22
Atomic mass	47.9
Crystalline structure α-β	Cubic compact hexagon, centered body
Density, kg/dm^3	10.6
Atomic volume, cm^3/atom*g	1668 ± 5
Melting temperature, $^\circ$C	1668 ± 5
Boiling temperature, $^\circ$C	3500 (estimate)

Compound: Oxidation number +4 is dominant in titanium chemistry, but +3-oxidation state compounds are very common. Due to this high valence state, many titanium compounds have a high tendency to form covalent bonds. Sapphire and ruby are related to the star properties of titanium dioxide impurities. Titan is also made of this substance. Barium titanium has piezoelectric properties, so it can be used as a converter in the conversion of sound to electricity. Titanium ester is formed by the reaction of alcohol and titanium tetrachloride and is used as a waterproof material [18,19].

Titanium nitrogen (TiN) is often used to cover cutting tools, such as drills (Figure 2.5).

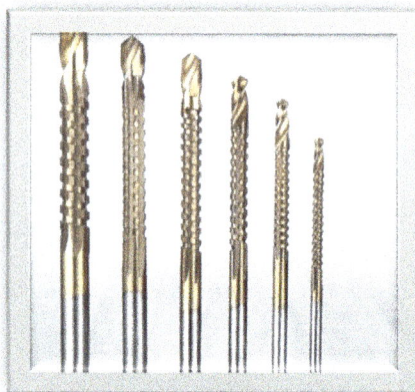

Figure 2.5. Drill covered with a layer of titanium [20].

Titanium tetrachloride is a colorless liquid that is used as an intermediate in the processing of titanium dioxide for paint. Titanium also forms a chloride with a lower valence, namely titanium trichloride, which is used as a reducing agent.

2.2. Classification, influence of alloying elements on titanium properties

According to the microstructure, titanium alloys are single-phase and diphasic. The alloying elements increase or decrease the temperature of $\alpha \rightarrow \beta$ phase transformations, thus varying both the α domain and the β domain of the alloy. This is how they differ [21]:

a) α-gene elements (Al, O, N, B, C);

b) isomorphic β-gene elements (V, Mo, Ta) that increase the β domain;

c) neutral isomorphic elements (Zn, Sn) that do not modify any domain.

The classification of titanium alloys used in practice can be done based on several criteria: processing, property (especially breaking strength), structure and fields of use [22, 23].

a) According to the processing method, titanium alloys are divided into two main categories: deformable alloys and alloys for foundries. Unlike other metals in the case of titanium-based alloys there are no essential differences between the compositions of those two types of materials;

b) According to the properties, they are divided into alloys with high plasticity and medium resistance, sufficiently plastic and with high resistance, with very good corrosion resistance, with special mechanical properties at temperatures below $0°C$, super plastic, superconducting, amorphous, with memory and so on.

c) According to the fields of use, these metallic materials can be grouped in the following categories: alloys for welded constructions, alloys used in the construction of motor vehicles, aviation, space technology, chemical industry, naval, in cryogenics and with special destinations.

d) According to the structure, the alloys are divided into three categories: α (alloying elements dissolve in α titanium); β (alloying components balance out this design at encompassing temperature) and $\alpha + \beta$ (biphasic) thus, biphasic combinations are separated into three subgroups: normal biphasic amalgams, pseudo α and pseudo β. The pseudo α will be α combinations that contain limited quantities of β stage in the construction (3… 10%). They hold every one of the fundamental qualities of α combinations, however the presence of the β stage adds to the improvement of innovative and mechanical properties. Pseudo β amalgams are β compounds that contain in the design modest quantities of α stage that satisfies the job of hardener in the solidifying and maturing process.

There are four types of non-alloy titanium for surgical implant applications (Table 2.3). The impurities contained distinguish them; oxygen, iron and nitrogen must be carefully controlled. Oxygen, in particular, has a great influence on the ductility and strength of titanium.

Table 2.3. Chemical composition of pure titanium (F67, ASTM, 2000) [24].

Element	Grade 1	Grade 2	Grade 3	Grade 4
N	0.03	0.03	0.05	0.05
C	0.10	0.10	0.10	0.10
H	0.015	0.015	0.015	0.015
Fe	0.20	0.30	0.30	0.50
O	0.18	0.25	0.35	0.40
Ti	balance	balance	balance	balance

Ti6Al4V

The first Ti alloy that combines Ti-specific biocompatibility properties with mechanical properties at least as good as those of conventional materials is the TiAl6V4 alloy. It is one of the most used titanium alloys in the medical field, Ti-6Al-4V is composed of 6% aluminum and 4% vanadium. The Ti-6Al-4 V alloy has low wear resistance, high modulus of elasticity (approximately 4-10 times higher than human bone) and low shear strength [25-27].

The use of this alloy in implantology involves risks of toxic reactions due to the presence in the composition of vanadium and aluminum. Vanadium has a high cytotoxicity, and aluminum can even induce senile dementia.

Table 2.4. Chemical composition of TI6A14V alloys [4,11].

Element	Forging processing (F136, F620)	Casting (F1108)	Covering (F1580)
N	0.05	0.05	0.05
C	0.08	0.10	0.08
H	0.012	0.015	0.015
Fe	0.25	0.30	0.30
O	0.13	0.20	0.20
Cu	-	-	0.10
Sn	-	-	0.10
Al	5.5-6.5	5.5-6.75	5.5-6.75
V	3.5	3.5-4.5	3.5-4.5
Ti	4.5	balance	

Ti alloys used in implants are not sensitive to the phenomenon of corrosion under load as in the case of stainless steels [28].

The chemical composition of the TiAl6V4 alloy is: Nitrogen 0.05%; Carbon 0.08%; Hydrogen 0.0125; Iron 0.25%; Oxygen 0.13; Titan the rest. The main alloying element is Al (5.5 - 6.5%) and vanadium (3.5 - 4.5%), the chemical composition is shown in table 2.4. The main elements that make up this alloy are aluminum (5, 5-6.5% by weight) and vanadium (3.5-4.5% by weight). There are other titanium alloys used in the manufacture of implants (Table 2.4 for their chemical composition) [29].

Another Ti6A14V alloy (F1472) is very similar to the F136 alloy (Table 2.5). All are in% maximum premises.

Table 2.5. Chemical composition of processed Ti alloys [4,29].

Element	Ti8Al7Nb (F1295)	Ti13Nb13Zr (F1713)	Ti121Mo6Zr2Fe (F1813)
N	0.05	0.05	0.05
C	0.08	0.08	0.05
H	0.009	0.012	0.020
Fe	0.25	0.25	1.5-2.5
O	0.20	0.15	0.08-0.28
Ta	0.50	-	-
Al	5.5-6.5	-	5.5-6.75
Zr	-	12.5-14	5.0-7.0
Nb	6.5-7.5	12.5-14	-
Mo	-	-	10.0-13.0
Ti		balance	

Titanium is an allotropic substance that is found in a very narrow (compact) hexagonal structure (α-Ti) up to 882°C and in a centered cubic structure (β-Ti) above this temperature. The addition of other alloying elements along with titanium leads to a series of properties:

1. Aluminum tends to stabilize the α phase, ie to increase the transformation temperature from the α to β phase.

2. Vanadium stabilizes the β phase by lowering the transformation temperature from α to β (Figure 2.6). α alloys have a single-phase microstructure, characterized by the ability to be welded. The stabilizing capacity of these aluminum-rich substances in this group of alloys

is completed in excellent hardness and oxidation resistance at high temperatures (300-600°C). These alloys cannot be heat treated for hardening purposes because they are single phase.

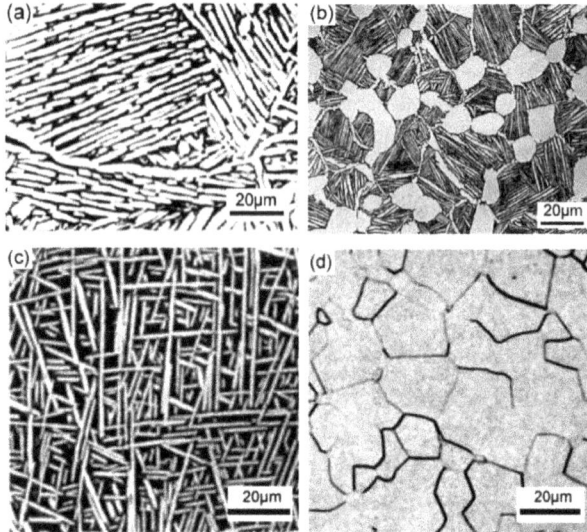

Figure 2.6. Typical microstructures of α + β titanium alloys: (a) Widmanstätten; (b) duplex microstructure; (c) basket-weave microstructure; (d) equiaxed structure [30].

The addition of a controlled amount of β stabilizing elements makes the upper β phase resistant to temperature transformations, which is achieved in a biphasic system. B-phase precipitates will appear after heat treatment at the tempering temperature of the solid solution, followed by aging at a certain low temperature. The aging cycle causes the precipitation of fine particles α from the metastable β that intersect a structure that is much more resistant than the annealed structure α-β. A higher percentage of β stabilizers (13% V weight in Ti13V11Cr3Al alloy) gives rise to a predominantly β microstructure that can be hardened by heat treatment [30].

The mechanical properties of commercial pure titanium and Ti6Al4V alloy are given in table 2.7. The coefficient of elasticity of these materials is 110 GPa, which is half the value of Co-Cr alloys. Also, can observe the high content of impurities which induces a high strength and a low ductility. The strength of Ti alloys is similar to that of type 316 stainless steel or Co-based alloys, as can be seen in tables 2.6 and 2.8.

Advanced Metallic Biomaterials   Materials Research Forum LLC
Materials Research Foundations **118** (2022)   https://doi.org/10.21741/9781644901779

Table 2.6. Mechanical properties of pure Ti (F67, 1992) [31,32].

Properties	Grade 1	Grade 2	Grade 3	Grade 4
Tensile strength ksi (MPa)	35(240)	50(345)	65(450)	80(550)
Flow limit 0.2% ksi (MPa)	25(170)	40(275)	55(380)	70(485)
Elongation (%)	24	20	18	15
Throat (%)	30	30	30	25

*1 ksi=1.000 psi, 1 psi= 6.895 Pa

Table 2.7. Mechanical properties of Ti6Al4V alloy [31,32].

Properties	Processing (F136)	Casting (F1108)
Tensile strength ksi (MPa)	125(860)	125(860)
Flow limit 0.2% ksi (MPa)	115(795)	110(758)
Elongation (%)	10 min	8 min
Throat (%)	20 min	14 min

*1 ksi=1.000 psi, 1 psi= 6.895 Pa. Forging of Ti6A14V alloy (F620) does not require mechanical properties

Table 2.8. Mechanical properties of processed Ti alloys [31,32].

Condition	Breaking strength, ksi, (MPa)	Flow limit (0,2% balance), ksi (MPa)	Elongation %	Surface reduction min,%
Ti8Al7Nb(F1295)	130,5(900)	116(800)	10	25
Aging	125(860)	105(725)	8	15
Hardened	80(550)	50(345)	15	30
Untempered	80(550)	50(345)	8	15
Ti121MoZr2Fe(F1813)	135(931,5)	130(897)	12	30

However, when compared in terms of specific strength (strength / strength). density), the titanium alloy excels over all other materials used in making implants. Titanium has a low edge strength, making it undesirable for bone screw implants, bone plates and other similar applications. It also tends to degrade or break when it comes in contact with itself or another metal.

Alloys with special properties - NiTi (Nitinol)

Titanium and nickel-based alloys have an unusual property, namely that if they are deformed below the polymorphic transformation temperature, they return to their original shape with increasing temperature. One of the best-known alloys based on titanium and nickel is the alloy Nitinol-55, which has in its composition the following elements: Ni and Ti in a proportion of 50-55%, as well as Co, Cr, Mn, and Fe [34].

The titanium-nickel alloy exhibits a number of quality properties such as good low temperature ductility, good biocompatibility, corrosion resistance, mechanical load resistance as well as the property of converting heat energy into mechanical energy. It is used in dentistry (dental implants), reconstructive surgery (cranial plaques), cardiac surgery (artificial heart) and orthopedics (clamps and fracture fixing screws) [35].

In addition, NiTi alloy has two important mechanical characteristics similar to natural biomaterials, such as bone (physiological compatibility).

First, NiTi alloy has a high degree of elastic recovery of up to 8%, which is close to that of bone (2%), while the reversal of specific deformations of stainless steel is only 0.5%. Second, the NiTi alloy has a lower modulus of elasticity of up to 48 GPa, which is close to that of bone (below 20GPa), while the modulus of elasticity of stainless steel can reach 193 GPa [36].

These properties recommend NiTi alloy as an ideal one, especially for orthopedic and orthodontic surgery. These properties of nitinol certainly make it the closest mechanical metal to biological materials and accelerate bone growth, improved adhesion to surrounding tissues, rapid cell regeneration, accelerate the healing of bone fractures, reduce healing time.

Permeable NiTi shape memory combination with a cell structure like a portion of the biomaterials, like bone, has been perceived as a promising biomaterial for the formation of counterfeit bones or dental roots. Contrasted with the properties of standard NiTi amalgam, the properties of the permeable NiTi combination can be effortlessly changed to relate to those of the bones by getting various porosities and distinctive pore sizes by controlling the blend conditions.

Besides, its permeable construction can permit the tissues of the human body to develop inside and body liquids to be moved through interconnected pores, consequently speeding up the recuperating system.

NiTi permeable combinations with three-dimensional interconnected pores have an isotropic permeable construction. The overall porosity of NiTi combinations can reach 57.3 vol.%, And the size of most pores is around 200–500 microns. The permeable combination with NiTi shape memory has a high compressive strength (208 MPa), a worth higher than that of a conservative human bone and can meet the prerequisites of hard tissue inserts for joint inserts [37].

The porous nitinol sponge maintains superelastic and shape memory properties, has a low modulus of elasticity, accelerates bone growth and provides improved adhesion to surrounding tissues. Applications using porous nitinol are very current as it further enhances the benefits of this material, especially bone growth.

From the point of view of the exigencies regarding the materials used for the hard tissue implants, a biomaterial characterized by a modulus of longitudinal elasticity reduced in value is desirable. An implant with a high Young's modulus value can cause high blood pressure concentrations that can weaken the bone and damage the implant / bone interface [38, 39].

The porous NiTi alloy is suitable for medical applications. Nitinol implants have been present in dentistry, orthopedics, cardiology and other medical fields, with a large number of temporary and permanent implants reported in Japan, Germany, China and Russia since 1980.

The best-known orthopedic applications in shape memory alloys made of Nitinol are staples for fibular fracture, for metacarpal fractures, for degenerative arthritis of the carpometacarpals and for hallux-valgus. Several types of shape-memory orthopedic implants with shape memory are used to accelerate the healing process of bone fractures.

As disadvantages, NiTi alloys can have allergenic effects due to nickel.

Experimental alloy (in vitro) Ti-Mo

Molybdenum (Mo) is a component with a lower level of poisonousness than Co, Ni, Cr and is a β-balancing out component. Studies in the field have shown that titanium amalgam molybdenum in rates between 15-20% can impact the decrease of the modulus of flexibility by crediting sufficient mechanical properties [36].

Ti-Mo-based compounds alloyed with different biocompatible components showed unrivaled mechanical properties, for example, high rigidity and a much lower modulus of flexibility, near that of human bone, contrasted with other traditional biomaterials. Studies have shown that $\alpha + \beta$ alloys have satisfactory characteristics, especially with superior mechanical properties, as well as increased resistance of α alloys, both to corrosion and oxidation. On the other hand, β-type alloys, due to the stabilizing elements Mo, Ta and Nb, have the advantage of increasing mechanical strength and a modulus of elasticity close to that of human bone, important aspects regarding the long-term use of biomaterials in the field medical [40,41].

An important aspect is that these alloying elements fall into the category of nontoxic elements, giving them the advantage of being used for applications in implantology.

Oliveira and his collaborators subjected to electrochemical testing in Ringer's solution, three types of biocompatible alloys: C.P. Ti, Ti6Mo and Ti15Mo at different time intervals. The electrochemical behavior showed that there are no major changes on the surface of the samples after 24 hours immersed in the solution used, forming layers of titanium oxides of surface thickness. The authors stated that the alloy with up to 15% molybdenum leads to the improvement of the corrosive protection characteristics, and the concentration of over 20%, leads to the inverse effect of the corrosive protection [42].

José Roberto Severino Martins Júnior and his collaborators studied the Ti15Mo alloy and the pure commercial titanium alloy, performing a cytotoxicity test. These alloys do not show a high degree of toxicity, respectively they are biocompatible with human tissue. The authors also studied cell morphology, indicating that the material did not cause any aggression and did not cause any change in cell morphology or its adhesion to the surface of the material [11].

Ti-Zr alloy

Zirconium (Zr) is a neutral component when broken up in Ti. Zirconium has a construction and substance properties like those of titanium. Along these lines, they were perceived as non-poisonous and non-unfavorably susceptible.

Zirconium is a progress metal with a nuclear number of 40 and a nuclear load of 91.22 amu, has amazingly high softening focuses (1857 ° C) and edge of boiling over (4409 ° C). Zirconium has a high consumption obstruction, like that of titanium and is biocompatible, as both metal surfaces structure a steady oxide layer on their surface. In any case, zirconium couldn't be utilized in dentistry in its unadulterated structure [43,44].

The binary alloy made out of a combination of titanium and zirconium (TiZr) was presented as an elective material for dental inserts.

It has been found that Ti alloys containing 50% Zr have higher values of tensile strength and hardness than Ti-6Al-4V.

Alloys with special properties

In this category can be included all titanium-based metallic materials that are widely used in cutting-edge technologies, among which can be mentioned: superconducting alloys, alloys for the electronics industry, super plastic alloys, amorphous alloys, memory alloys [45].

Advanced Metallic Biomaterials   Materials Research Forum LLC
Materials Research Foundations **118** (2022)   https://doi.org/10.21741/9781644901779

Memory alloys are metallic materials that under certain conditions have the property of changing their dimensions. After the suppression of external factors, they have the ability to return to their original form.

Superplastic alloys are metallic materials that in certain temperature ranges have values for elongations greater than 100%. These metallic materials allow, if the processing is performed at a temperature included in this field, to obtain by modeling products with extremely varied configuration. In this category can be included Ti alloys with α and $\alpha + \beta$ structure, two being the most studied compositions: 6% Al, 4% F at 950°C, respectively 5% Al, 2.5% Sn, rest at% Ti at 900°C. Superconducting alloys are materials that at low temperatures become superconducting.

Amorphous alloys are metallic materials with an amorphous structure (metal glass), which are obtained by cooling rates with special properties, especially regarding the ability to concentrate light sources. These alloys have already found applications in the electronics industry, for deposition of thin layers in solar batteries [46].

The influence of alloying elements on titanium alloys

Zircon and hafnium, titanium-like metals - form continuous series of solid solutions with both allotropic modifications, shown in the equilibrium diagrams, Ti-Hf, in the temperature range corresponding to the βTi phase, the martensitic transformation of the β phase into the α 'phase occurs as in pure titanium. Only when hardening alloys containing more than 20% Zr can a certain amount of β-phase be fixed. These two alloying elements are used because they increase the refractoriness and creep strength of the metal. In addition, ziconium in Ti-Al alloys leads to improved plasticity as do β-stabilizing elements [47].

Mobibdenum, vanadium, niobium, tantalum or rhenium form with titanium binary alloy systems with state diagrams close to zirconium, as they are part of the 5th group of its system. Mendeleev, and the diameters of their atoms differ from that of titanium by less than 10 .... 15%, having an isomophic crystal lattice with β phase. Unlike zirconium and hafnium, their αTi solubility is limited due to deferences in the construction of crystal lattice.

When hardening in the β range, this structure can be fixed in a proportion of 100% in binary alloys containing 10% Mo, 15% V, 36% Nb, 60% Ta. If these elements are added in small quantities, during the hardening the process of martensitic transformation of the β phase into the α 'phase takes place. However, if the hardening is performed at range temperatures ($\alpha + \beta$), the β phase can be fixed even at lower contents of the titanium-based alloying elements, contributing to the simultaneous improvement of the different properties, including the corrosion resistance. According to their solubility n αTi they can be arranged in the following order: 0.8% Mo, 1 ... 3.5% V, 3 ... 4% Nb, 6 .... 12.5% Ta [48, 49].

The study of the equilibrium diagrams shows the lack of intermetallic phases, eutectoid and peritectic reactions, which determines their superiority over other alloying elements.

Chromium being isomorphic with βTi forms the continuous series of solid solutions. The solubility of αTi is low (below 0.5%), and the phase undergoes a eutectoid transformation with the separation of the α phase and the $TiCr_2$ intermetallic compound [50].

By hardening the β domain in alloys containing more than 7% Cr this phase can be fixed. Its eutectoid decomposition occurs during prolonged heating of the hardened alloy in the temperature range 350 C - transformation temperatures.

Alloys based on the Ti - Cr system are characterized by very good mechanical properties; chromium being considered along with molybdenum as the best alloying element. Lately, its significance has decreased because of the appearance of fragility due to eutectoid transformation.

Manganese, iron, cobalt, nickel, tin, magnesium, gold, pallium, silver and silicon form equilibrium diagrams with eutectic and eutectoid reactions.

Of these elements, manganese has the greatest importance, as it is accessible and cheap. Similar to chromium, it can replace elements in the β-stabilizing group in a number of alloys. The speed of the eutectoid reaction is much lower than in the case of chromium alloys, which limits the appearance of fragility.

Iron - the next important alloying element after manganese is used as chromium in complex titanium alloys having a positive influence on mechanical properties.

The alloys of the Ti - Cu system was studied from the point of view of the particularity of the structure and of the possibility of hardening by heat treatment. In these alloys, it is not possible to fix the β phase by hardening, they having at ambient temperature an α structure and a certain amount of intermetallic phase [51].

The influence of tin on the properties of titanium has been studied by various researchers. There is no opinion on the construction of the equilibrium diagram. The latest works indicate the formation of a eutectoid. There is an insignificant difference between the eutectoid reaction temperature and the allotropic transformation temperature (17°C), as well as between the content of the saturated solid solution α and the eutectoid cancer concentration (1%) [52, 53].

Tin is one of the best alloying elements for α alloys. In high concentrations (over 10%), it has a behavior similar to that of copper forming alloys with intermetallic compounds.

The mechanism of eutectoid transformation in the Ti - Sn system has been little studied, because the β phase is not fixed by hardening, and the eutectoid transformation temperature is high (865 C) and it can be assumed that the reaction rate is very high.

Silver has a fairly high solubility in titanium α (maximum 12.5% at 800 °C), and can form intermetallic phases with high plasma, which causes the alloys in this system to be used as soldering metal materials.

The addition of silicon (in the order of tenths of a percent) leads to an increase in the refractoriness of steel alloys, especially those with αxβ structure. This favorable action of silicon is explained by the appearance in the structure of titanium silicides or alloying elements [54].

Boron, cerium, lanthanum and scandium form binary systems with titanium, whose state diagram shows eutect and territorial reactions. The only one used so far in the structure of titanium alloys is boron, which favors the finishing of the structure.

Aluminum is one of the few alloying elements that leads to an increase in the allotropic transformation temperature of titanium [55, 56].

The influence of thermomechanical treatment on the microstructure and properties of Ti alloys

Construction engineers include titanium alloys in the category of alloys with an inappropriate behavior from their point of view, due to the very large variations of the characteristics and behavior of the semi-finished products subjected to an identical forging or heat treatment regime. These variations occur because the properties of titanium alloys are very sensitive to characteristics such as texture, grain size, β phase transformation.

Stamping can be performed only in the single-phase domain β, or α, or in the biphasic one (α + β), or (because the molding temperature is not usually maintained at the same level) in different phase domains corresponding to the different stages of the molding process. As the subsequent heat treatment can be performed in any indicated potential phase range, a large number of heat treatment regimes can be applied [57].

But during the different stages of thermomechanical treatment a very rigorous control is necessary, oriented not only to ensure an advanced microstructural uniformity in sections of different thicknesses, which will implicitly lead to a minimal variation of the properties of the alloys [5].

Stamping in the β domain can be achieved isothermally, at temperatures corresponding to the domain of existence of the β phase, with the completion of the molding in the domain (α + β), through the normal cooling process during this operation. Thus, in the molding process in the β domain, more than 75% of the compression is actually performed in the (α + β) domain [58].

The main advantage of molding in the β range is that the deformation process can be performed in a wider range of temperatures than in the case of molding in the range (α + β) when the permissible temperature range for molding is much narrower. This advantage allows molding to be performed in a smaller number of compressions and reheating to obtain the final shape. In addition, due to much lower values of the flow resistances, it is possible to stamp parts with a much smaller material addition to the final dimensions, so with an economical material utility.

Related to the fact that the leakage of the β phase is largely sensitive to the deformation speed, there is the advantage of the possibility of molding at low deformation speeds, a process easier to perform in isothermal conditions, in which the part can be molded in one operation [5].

Stamping in the (α + β) domain can be performed in the initial stages in the β domain, but to obtain a homogeneous biphasic (α + β) structure it is important that all areas of the semi-finished product are deformed in this field, otherwise the areas of dead metal will have an acicular microstructure characteristic of molding or heat processing in the β domain.

In the isothermal molding process in the domain (α + β) the supraplasticity properties of the fine microstructure (α + β) can be used. Most industrial titanium alloys with structure (α + β) can be brought into the supraplastic state by hot deformation with high compression rates achieved in the field (α + β): thus, the coarse phase mixture stabilizes the size of the gasket, which, by subsequent recrystallization will decrease [59, 60].

For example, in the case of Ti-6Al-4V alloy, it has been shown that by the tests of stretching them with deformation speeds of 10-5-10-3 in the temperature range 900-950°C, the size of the sensitivity coefficient of the deformation speed is reached of 0.85 and elongation ~ 1000% [5].

2.3. Applications in medicine

The first tests of titanium in medical implants date back to the 1930s. The light weight (4.5 g / cm^3) as well as the very good mechano-chemical properties of titanium, make it a widely used material for orthopedic implants. There are four categories of titanium used in medical applications. The differences between them are given by impurities such as: oxygen, iron and nitrogen. In particular, oxygen has a good influence on the ductility and mechanical strength. In addition to the components presented above, other components are also used, such as: hydrogen and carbon (0.015% and 0.1%, respectively). Titanium also has a very high resistance to corrosion, due to the formation of a layer of titanium oxide (TiO_2) on its surface. This film accelerates the process of osseointegration, a process by which bone tissue adheres to the surface of the implant without the appearance of chronic inflammation [61].

Titanium is non-poisonous even in enormous amounts and plays no normal part in the human body. An expected 0.8 milligrams of titanium are ingested by people each day, however most pass through the body without being assimilated. It does, notwithstanding, tend to bio-collect in silicon dioxide-containing tissues. An obscure plant framework might utilize the metal to invigorate sugar creation and empower development. This could clarify why most plants have one section for each million (ppm) of titanium, food plants have 2 ppm, and horsetails and brambles have up to 80 ppm [62].

As a powder or as metal filings, titanium represents a huge danger of fire and, whenever warmed in air, a danger of blast. Water and carbon dioxide strategies for stifling flames are inadequate on consuming titanium; D-type putting out fires specialists as anhydrous powder ought to be utilized all things being equal. When used in the manufacture or handling of chlorine, precautions should be taken for the use of titanium only in places where it will not be exposed to anhydrous chlorine gas, which may result in a titanium / chlorine fire. There is a risk of fire even when used in hydrated chlorine due to the unexpected drying of the gas, caused by extreme weather conditions [63].

Titanium can burst into flames when a new, non-oxidized surface interacts with fluid oxygen. These surfaces can seem when the oxidized ones are hit with a weighty item, or when a mechanical pressure makes a break show up. This is a potential limit in the utilization of titanium in fluid oxygen frameworks, like those found in the aeronautic trade.

Advantages of titanium for implantology [64-66]:

1). Titanium is a reactive material. This means that in air, water or in contact with any electrolyte it is spontaneously covered with a layer of oxide. This oxide is one of the hardest known, forming a dense film, which protects the metal from chemical attack including the aggressive attack of liquids and organisms. Titanium is inert in tissues. The dense oxide layer in contact with the tissues is practically insoluble. In particular, no ions are released that could react with organic molecules.

2). Titanium has good mechanical properties. Its hardness is very close to that of stainless steel, used for load-bearing surgical implants. Titanium is much harder than cortical bone or dentin, allowing dental implants to be made

thin shape, which are able to withstand heavy loads. It is very important that the metal is strong and malleable, which makes it insensitive to a shock load.

3). Titanium does not behave passively in tissues and bone, the bone grows on the rough surface and binds to the metal.

Disadvantages of titanium include relatively low shear strength, low wear resistance and manufacturing difficulties [67, 68].

Titanium and nickel-based alloys have an unusual capacity, namely, if they are deformed below the polymorphic transformation temperature, they return to their original shape with increasing temperature. One of the most well-known alloys based on titanium and nickel is the alloy Nitinol-55, which has in its composition the following elements: Ni and Ti in a proportion of 50-55% as well as Co, Cr, Mn and Fe. This type of alloy exhibits a number of quality properties such as good low temperature ductility, good biocompatibility, corrosion resistance, mechanical load resistance as well as the property of converting caloric energy into mechanical energy. It is used in dentistry (dental implants), non-constructive surgery (cranial plaques), cardiac surgery (artificial heart) and orthopedics (clamps and screws for fixing fractures).

Approved and used titanium biomedical alloys are: Ti-Al-V, Ti-Al-Mo, Ti-Al-Cr, Ti-Al-Cr-Co [69].

Titanium-based materials used in implantology are evaluated from two points of view: physical-mechanical and biological.

From a physical-mechanical point of view:

1. melting point - 1600°C, ultra fast sterilization at 300°C.

2. strength, rigidity - implants, titanium milling cutters are made of a single bar by mechanical processing, which gives it maximum strength. Implants, milling cutters do not deform when applying mounting forces, milling, or mastication biomechanics, even thin implants withstand heavy loads. The strength Ti is comparable to that of stainless steel. Hardness is much higher than cortical bone and dentin. It is malleable, which makes it resistant to shock demands.

3. cathodic effect - It acts as a cathode, attracting calcium ions around it, favoring the appearance of hydroxyapatite nucleus.

4. neutral pH -7- of Ti oxide.

5. thermal conductivity - low.

6. electrical resistance - increased.

7. weight, density - low. It is located between heavy and light metals, closer to light ones, so the weight exerted by the implant on the surrounding cells is reduced.

From a biological point of view - of the reaction of tissues to Ti:

1. corrosion resistance - it is a reactive material - in water, air, or any other electrolyte it is spontaneously covered with a layer of titanium oxide. This oxide is one of the most resistant minerals known to form a dense, compact, stable, insoluble film that protects Ti from chemical attack, including the aggressive one produced by body fluids. The oxide gives it corrosion resistance.

2. amagnetism - have no magnetic effect; it does not produce a magnetic field that disrupts the activity of the surrounding cells.

3. regenerative activity, therapeutic-healing qualities of Ti oxide, being used in dermatological treatments.

4. biological compatibility - the oxides on the implant surface being very adherent and insoluble prevent the release and direct contact between potentially harmful metal ions and tissues.

Daily intake of Ti is important 40% of the amount ingested daily, which is 300 g and is metabolized. The amount of Ti resulting from the oxidation of an implant inserted into the bone is 10,000 times lower than that metabolized. Thus, the presence of a Ti implant is irrelevant to the total amount of Ti in the body, there are no systemic reactions, allergies, deposits in organs. Only inter-, intra-cellular impregnations were demonstrated after milling, but without affecting the cell functions.

5. osseointegration-between the Ti implant and the surrounding bone, a solid connection is established by increasing the bone on the rough surface of the metal and connecting it, achieving an ankylosing, mechanical, rigid anchorage, stabilizing the endo-bone implant.

According to some authors, this ankylosing spondylitis is equivalent to osseointegration. Branemark demonstrates that the Ti oxide layer that covers the implant establishes a bivalent bond at the molecular level with the elements of the bone tissue.

From a histological point of view, osseointegration is materialized by the presence of regenerated bone in the immediate vicinity of the metal surface.

Studies of the implant-bone contact structure, of the bone-implant interface, were performed using X-rays, scanned micrography, highlighting the lamellar bone with its characteristic gaps penetrating the porous surface of the Ti plasma, the bone approaching less than 0.5m from the surface of the metal, space too small for the presence of any tissue organized between bone and metal [70,71].

The bone-implant interface being a diffuse interface, in order to obtain a surface on the implant with a higher anchoring power in the bone, plasma coating is used. It is made with the help of an inert gas in a voltaic arc at a very high temperature resulting in plasma and hybrid coating material -Ti, which is projected on the surface of the implant. The result is the appearance of a layer of 20-30 m thickness and 15 m roughness that creates a rough surface with round shapes, highly porous, having an effective surface 12 times larger than an implant [72].

Ti-coating hybrid, like Ti, in contact with air, water, or any other electrolyte, forms Ti oxides that give it the same chemical, biological compatibility properties as a solid Ti implant.

According to some researchers, the connection between the sheath and the implant is not strong enough to oppose the tensile forces between the implant and the bone, leading to its rupture. According to other researchers, the eventual fracture of the implant occurs before the separation of the plasma coating from the implant.

Corrosion products of titanium: titanium dissolves in the form of trivalent ions, which immediately binds to organic molecules forming new very stable metal complexes, concentrating especially at the site of implantation. At a distance, titanium is found in the spleen and lungs, being absent in the liver and kidneys. Allergic properties of titanium have not been described so far, but they certainly form very stable metal complex proteins with proteins. In the presence of common metals, titanium induces a hypersensitivity reaction, which requires its use in pure form, without the presence of other common metals in the vicinity [73].

A study of the response of soft tissue to titanium shows that in the short term of 6 months, titanium introduced into muscle mass is well tolerated with the appearance of the classic fibrous capsule of variable thickness, similar to that which occurs in stainless steel. Titanium can accumulate in the surrounding tissues in 2 forms:

- Type A particles, negative to the Perl reaction;

- Type B particles, positive to the Perl reaction.

The surrounding tissue will therefore contain areas of necrosis and titanium particles included in the fibrous or granular tissue. Type B particles are found in cytoplasmic inclusions, macrophages and fibrocytes, which are viable alive without the occurrence of inflammatory reactions.

A study of the response of bone tissue to the titanium implant shows that the study of the ultrastructure of the bone-implant interface, under the scanning electron microscope, highlighted three distinct areas, as follows:

- The glycoprotein area in direct contact with the bone and the surface of the implant;

- Area of disordered and calcified collagen fibers,

- The area of ordered and calcified collagen fibers.

These areas start from the implant to the bone and their thickness is dependent on the biocompatibility of the biomaterial.

The protein layer is composed of glycoproteins associated with a network of hyaluronic acid which is the biological fluid that holds the fibers and cells together. This protein layer is in contact with the implant and has a variable thickness of 200-400 A. If in titanium and in aluminum and zirconium ceramics this layer decreases in parallel with the physiological bone remodeling in close connection with the gradual prosthetic loading, in the rest it enlarges by fibrosing. So, the shrinking proteoglycan layer is an indication of the osseoacceptance of the implant in the bone.

Layer 2 of disordered collagen fibers forms a three-dimensional peri-implant network that adheres to the proteoglycan layer at the implant surface

The third layer of ordered and calcified collagen fibers is at a distance of 1000-2000 A from the implant surface.

In conclusion, we can say that:

- Titanium and ceramics are the materials of choice in implantology, titanium being used as a biomaterial for over 30 years with very good results proven both experimentally and clinically, due to its properties, represented by: amagnetism, biological immunity, therapeutic action, resistance, homogeneity and purity, low specific gravity.

- A material is biocompatible if it produces only desired or tolerated reactions in a living organism or does not produce unwanted tissue reactions.

Titanium and its alloys are used to make orthopedic and dental implants due to the fact that its mechanical properties are similar to those of bone tissue.

Here are some examples of titanium implants:

Shape memory plates (Figure 2.7) Are used where no molding can be applied to the surrounding area, for example, facial area, nose, jaw, eye area. They are placed on the fracture and are fixed with screws, preserving the original alignment of the bones and allowing cell regeneration. Due to the shape memory effect, when heated, these plates tend to return to their original shape, exerting a constant force that ensures the union of the fracture fragments, helping in the recovery process [74].

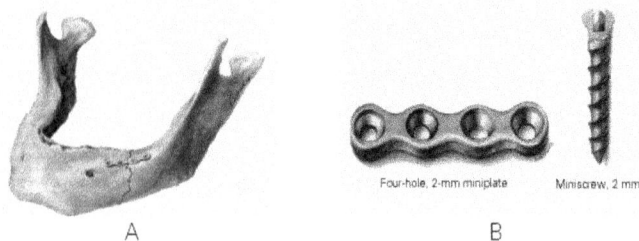

Four-hole, 2-mm miniplate   Miniscrew, 2 mm

A   B

Figure 2.7. Shape memory plates: a) mounted plate, b) examples of miniplate and miniscrew [75].

Harrington rods (Figure 2.8) made of Ni-Ti have a much-simplified construction compared to classic devices, with steel hooks that attach to the spine to the two sides of the scoliotic curve. In addition, the classic rods gradually relax, both during and after the operation, so that after 10-15 days the stretching force of the spine decreases to approx. 30% of the initial value, which generally requires the performance of the second operation. For Ni-Ti rods, with AF ≈ 430°C, an external heating is applied after the relaxation period, which determines the return to the initial length, following an elongation of approximately 1 cm, restoring the correct stretching force of the vertebrae.

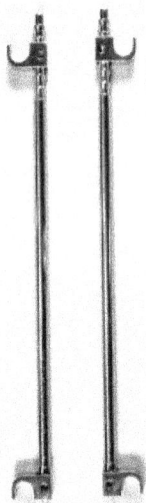

Figure 2.8. Harrington rods [76].

The use of wire (Figure 2.9) or circles to couple bone fragments is simple, but applicable to the Galeazzi fracture-dislocation case. They are found: Bone suture - in fractures of wide bones (back, pelvis). Holes are drilled in the bone, through which a wire is passed. Cerclage (circling) - in diaphyseal fractures in the clarinet beak. It must remain an exceptional method and only for long oblique and spiroid fractures and in the absence of a more adequate means of osteosynthesis.

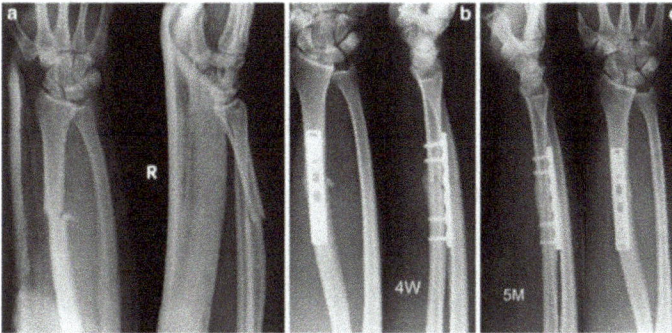

Figure 2.9. a) Galeazzi lesion of the right forearm. b) Radiograph after 4 weeks with elastic bridge plating with two empty holes above the fracture. A small fragment is left in the soft tissue without repositioning (no-touch technique). Visible callus formation. c) Callus resorption after fracture healing in progress even after 5 months [77].

Use of clamps, staples, screws or plates with osteosynthesis screws. These techniques are used especially for the bones in the vicinity of the joints.

Screwing (Figure 2.10) is done with screws and the screw pitch is very small for compact bones, large pitch for spongy bone (epiphyses).

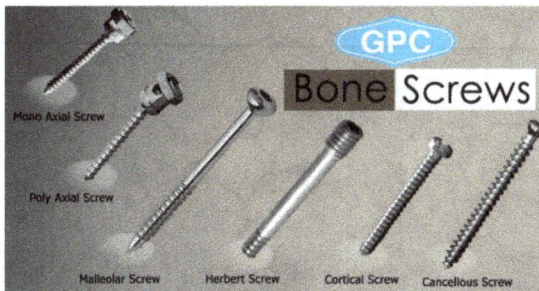

Figure 2.10. Types of Orthopedic Bone Screws [78].

Materials Research Forum LLC
https://doi.org/10.21741/9781644901779

Clamps or clips are used either to fix fracture fragments or to fix osteotomy fragments of small bones. It achieves a fixation and a relative stabilization of the bone fragments and also prevents the action of compaction forces, causing the delay of bone consolidation and the appearance of pseudarthrosis.

The screw plate is used for better compaction of the fracture edges by better fracture stability, thus allowing primary fracture healing by forming a minimal bone callus and providing increased strength. The stability of the implant-bone assembly is improved by the increased resistance to bending forces at the fracture site and the torsional forces applied to the plate are reduced by the friction between the fracture edges, increasing the life of the plate. However, the plates require a sufficient and obligatory length and the surgery for mounting the metal implant requires large incisions with large tissue damage, high blood loss, high tissue exposure, which increases the risk of infections, with high risks in their spread to bone (bone infections are incurable) and with unsightly scars [79, 80].

Nail-plate and blade-plate, (Figure 2.11) are used especially in the case of varicose and derotation femoral osteotomies. It is performed in the sub or intertrochanteric region. It has the same drawbacks as the screw plate, in addition it is very laborious to perform implantation with these materials, which leads to prolonged exposure time of the wound to the environment, leading to increased risk of infections, large tissue damage and high loss of blood. The percentage of pseudoarthrosis and erection damage before fracture consolidation is very high [81, 82].

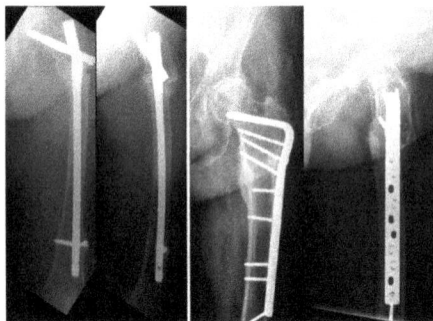

Figure 2.11. Fixing with blade-plate and nail-plate [83].

Centromedullary shaft is recommended for cases of communicative fractures, proximal or distal third fractures, for fixing diaphyseal fractures of long bones (femur, tibia, humerus) which limits their use. It achieves good focusing, but compaction is quite poor.

When blocking these rods by passing a proximal screw and a transverse one through the bone and the rod, there is a cancellation of the compaction forces and, therefore, the delay of consolidation, with the appearance of pseudarthrosis.

Centromodular brooches are devices used for endosseous fixation, where other ways of stabilizing the fracture are difficult to perform. The brooch variants have applications in the case of femoral neck fractures: Brose Steinmann and Knowles. Hansson brooches are used for fractures of the shoulder (Figure 2.12) and for neck necks, self-tapping brooches are used.

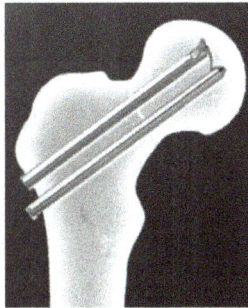

Figure 2.12. Hansson brooches [84].

External fixators (Figure 2.13) are recommended for open fractures with massive tissue destruction, ensuring instant fixation of traumatized bones. It is often the only solution in treating deficient bones or infected traumatized areas.

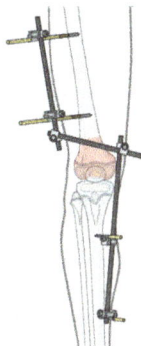

Figure 2.13. External fixators

We believe that orthopedics will become in the future the most important source of income among all medical branches, under the influence of the population explosion, accelerated technological development and globalization. has been more lucrative in the past, what the American Association of Orthopedic Surgery (AACO) calls the "decade of orthopedics" provides the greatest profit opportunities in health care [3].

References

[1] Leyens, C.; Peters, M. Titanium and Titanium alloys. Fundamentals and Applications, *Ed. Wiley-VCH*, **2003**, 423-431. https://doi.org/10.1002/3527602119

[2] Chen, Q. ; Thouas, G.A. Metallic implant biomaterials, *Materials Science and Engineering R* **2015**, *87*, 1–57. https://doi.org/10.1016/j.mser.2014.10.001

[3] Geetha, M.; Singh, A.K.; Asokamani, R.; Gogia, A.K. Ti based biomaterials, the ultimate choice for orthopaedic implants - A review, *Mater. Sci.* **2009**, *54*, 397-425. https://doi.org/10.1016/j.pmatsci.2008.06.004

[4] Lŭtjering, G.; Williams, J.C. Titanium-Second Edition, *Springer Science + Business Media*, Germany, **2000**.

[5] Niinomi, M. Mechanical properties of biomedical titanium alloys, *Mater. Sci. Eng., A*, **1998**, *243*, 231-236. https://doi.org/10.1016/S0921-5093(97)00806-X

[6] Lipşa, C.; Lipşa, D.S. Biomateriale-Curs pentruanul I, Iaşi, **2009**.

[7] Balaban, D.P.; Biomateriale, *Ed. Ovidius University Press*, Constanţa, **2005**.

[8] Information on https://www.worthpoint.com/worthopedia/titanium-metal-crystal-bar-element-1852804059. Available online: (accessed on September 10, 2021).

[9] Bunea, D.; Nocivin, A. Materiale biocompatibile, *Ed. şi Atelierele Tipografice Bren*, Bucureşti, **1998**.

[10] Popa, C.; Cândea, V.; Şimon V.; Lucaciu, D.; Rotaru, O. Ştiinţa biomaterialelor, *Ed. U.T. Press*, Cluj-Napoca, **2008**.

[11] Cercel (cas. Bălţatu), M.S. Contribuţii privind îmbunătăţirea proprietăţilor aliajelor de Ti-Mo destinate aplicaţiilor medicale – teză de doctorat, Iaşi, **2017**.

[12] Ţuculescu, S.U.; Bratu, E.; Lakatos, S. Titanul în stomatologie, *Ed. Signata*, Timişoara, **2001**.

[13] Information on https://www.britannica.com/science/titanium. Available online: (accessed on September 10, 2021).

[14]   Schwartz, A.J.; Kabo, M.J.; Eilber, F.C.; Eilber, F.R.; Eckardt, J.J. Cemented Distal Femoral Endoprostheses for Musculoskeletal Tumor: Improved Survival of Modular versus Custom Implants. *Clinical Orthopaedics and Related Research* **2010**, *468(8)*, 2198-2210. https://doi.org/10.1007/s11999-009-1197-8

[15]   Pourbaix, M. Atlas of Electrochemical Equilibria in Aqueous Solutions, 2nd ed.; NACE: Houston, TX, USA, **1974**.

[16]   Acevedo-Pena, P.; Vazquez-Arenas J.; Cabrera-Sierra, R.; Lartundo-Rojas, L.; Gonzalez, I.; Ti Anodization in Alkaline Electrolyte: The Relationship between Transport of Defects, Film Hydration and Composition. *Journal of The Electrochemical Society* **2013**, *160(6)*, C277-C284. https://doi.org/10.1149/2.063306jes

[17]   Dobrescu, M.; Dumitrescu, C.; Vasilescu, M. Titan şi aliaje de titan, *Ed. Printech*, Bucureşti, **2000**.

[18]   Steinemann, S.G.; Mausli, P.A.; Szmukler-Moncler, S.; Semlitsch, M.; Pohler, O.; Hintermann, H.E.; et al. Beta titanium in the 1990s. Warrendale, Pennsylvania: The Mineral, *Metals and Materials Society* **1993**, 2689–1696.

[19]   Collings, E.W. The Physical Metallurgy of Titanium Alloys, American Society for Metals, **1984**.

[20]   Information on https://www.accesorii-unelte.ro/product/set-6-burghie-din-titan/. Available online: (accessed on September 10, 2021).

[21]   Chelariu, R.; Bujoreanu, G.; Roman, C. Materiale metalice biocompatibile cu baza titan, *Ed. Politehnium*, Iaşi, **2006**.

[22]   Wang, K. The use of titanium for medical applications in the USA, *Mater. Sci. Eng. A*, **1996**, *223*, 134-137. https://doi.org/10.1016/0921-5093(96)10243-4

[23]   Bălţatu, M.S.; Vizureanu, P.; Istrate, B. Physical and structural characterization of Ti-based alloy, *International Journal of Modern Manufacturing Technologies* **2015**, *VII(2)*, 12-17.

[24]   Zhou, Y.L.; Luo, D.M. Microstructures and mechanical properties of Ti–Mo alloys cold-rolled and heat treated, *Materials characterization* **2011**, *62*, 931-937. https://doi.org/10.1016/j.matchar.2011.07.010

[25]   Valente, M.L.d.C.; de Castro, D.T.; Macedo, A.P.; Shimano, A.C.; dos Reis, A.C. Comparative analysis of stress in a new proposal of dental implants. *Mater. Sci. Eng. C* **2017**, *77*, 360–365. https://doi.org/10.1016/j.msec.2017.03.268

[26] Baltatu, M.S.; Tugui, C.A.; Perju, M.C.; Benchea, M.; Spataru, M.C.; Sandu, A.V.; Vizureanu, P. Biocompatible Titanium Alloys used in Medical Applications, *Revista de Chimie* **2019**, *70(4)*, 1302-1306. https://doi.org/10.37358/RC.19.4.7114

[27] Elias, C.N.; Oshida, Y.; Lima, J.H.; Muller, C.A. Relationship between surface properties (roughness, wettability and morphology) of titanium and dental implant removal torque. *J. Mech. Behav. Biomed. Mater.* **2008**, *1*, 234–242. https://doi.org/10.1016/j.jmbbm.2007.12.002

[28] Okazaki, Y.; Ito, Y.; Kyo, K.; Tateishi, T. Corrosion resistance and corrosion fatigue strength of new titanium alloys for medical implants without V and Al. *Mater. Sci. Eng. A* **1996**, *213*, 138–147. https://doi.org/10.1016/0921-5093(96)10247-1

[29] Niinomi, M. Recent research and development in titanium alloys for biomedical applications and healthcare goods. *Sci. Technol. Adv. Mater.* **2016**, *4*, 445–454. https://doi.org/10.1016/j.stam.2003.09.002

[30] Zhong, C.; Liu, J.; Zhao, T.; Schopphoven, T.; Fu, J.; Gasser, A.; Schleifenbaum, J.H. Laser Metal Deposition of Ti6Al4V—A Brief Review. *Appl. Sci.* **2020**, *10*, 764. https://doi.org/10.3390/app10030764

[31] Bombac, D.M.; Brojan, M.; Fajfar, P.; Kosel, F.; Turk, R. Review of materials in medical applications, *Materials and Geoenvironment* **2007**, *54(4)*, 471-499.

[32] Elias, C.N.; Lima, J.H.C.; Valiev, R.; Meyers, M.A. Biomedical applications of titanium and its alloys, *Biological Materials Science*, **2008**, 46- 49. https://doi.org/10.1007/s11837-008-0031-1

[33] Mihăilescu, C.; Mihăilescu, A.; Georgescu, N. Biomateriale de uz chirurgical, *Ed. Tehnopress*, Iaşi, **2011**.

[34] Roşu, R.A. Metode de obţinere şi de prelucrarea biomaterialelor pentru proteze umane - teză de doctorat, *Universitatea "Politehnica"*, Timişoara, **2008**.

[35] Davis, J.R. Handbook of materials for medical devices, *ASM International*, United States of America, **2003**.

[36] Bălţatu, M.S.; Vizureanu, P.; Ţierean, M.H.; Minciună, M.G.; Achiţei, D.C. Ti-Mo Alloys used in medical applications, *Advanced Materials Research*, **2015**, *1128*, 105-111.

[37] Park, Y.J.; Song, Y.H.; An, J.H.; Song, H.J.; Anusavice, K.J. Cytocompatibility of pure metals and experimental binary titanium alloys for implant materials. *Journal of dentistry* **2013**, *41*, 1251-1258. https://doi.org/10.1016/j.jdent.2013.09.003

[38] Uhm, S.H.; Song, D.H.; Kwon, J.S.; Lee, S.B.; Han, J.G.; Kim, K.M.; Kim, K.N. E-beam fabrication of antibacterial silver nanoparticles on diameter-controlled TiO_2 nanotubes for bio-implants. *Surf. Coat. Technol.* **2013**, *228*, S360–S366. https://doi.org/10.1016/j.surfcoat.2012.05.102

[39] Chen, P.C.; Hsieh, S.J.; Chen, C.C.; Zou, J. The microstructure and capacitance characterizations of anodic titanium based alloy oxide nanotube. *J. Nanomater* **2013**, 157494. https://doi.org/10.1155/2013/157494

[40] Catauro, M.; Bollino, F.; Papale, F. Preparation, characterization, and biological properties of organic-inorganic nanocomposite coatings on titanium substrates prepared by sol-gel. J. *Biomed. Mater. Res. A* **2014**, *102*, 392–399. https://doi.org/10.1002/jbm.a.34721

[41] Gordin, D.M.; Gloriant, T.; Nemtoi, G.; Chelariu, R.; Aelenei, N.; Guillou, A.; Ansel, D. Synthesis, structure and electrochemical behavior of a beta Ti-12Mo-5Ta alloy as new biomaterial, *Materials Letters* **2005**, *59*, 2936 – 2941. https://doi.org/10.1016/j.matlet.2004.09.063

[42] Pincovschi, E.; Florea, C.M. Compuşi anorganici biocompatibili cu aplicaţii în implantologie, *Ed. Printech*, Bucureşti, **1997**.

[43] Zhang, L.B.; Wang, K.Z.; Xu, L.J.; Xiao, S.L.; Yu-yong Chen., Y.Y. Effect of Nb addition on microstructure, mechanical properties and castability of type TiMo alloys, *Trans Nonferrous Met. Soc. China* **2015**, *25*, 2214-2220. https://doi.org/10.1016/S1003-6326(15)63834-1

[44] Zhan, Y.; Li, C.; Jiang, W. β-type Ti-10Mo-1.25Si-xZr biomaterials for applications in hard tissue replacements, *Materials Science and Engineering C* **2012**, *32*, 1664–1668. https://doi.org/10.1016/j.msec.2012.04.059

[45] Bălţatu, M.S.; Cimpoeşu, R.; Vizureanu, P.; Achitei, D.C.; Minciuna, M.G. Microstructural characterization of TiMoZrTa alloy, The Annals of „Dunărea de Jos" University of Galaţi, *Fascicle IX. Metallurgy and materials science* **2015**, *4*, 23-26.

[46] ***ASM Handbook, Alloy Phase Diagrams, vol.3.

[47] Bălţatu, M.S.; Vizureanu, P.; Geantă, V.; Nejneru, C.; Ţugui, C.A.; Focşăneanu, S.C. Obtaining and Mechanical Properties of Ti-Mo-Zr-Ta Alloys, *IOP Conference Series: Materials Science and Engineering* **2017**, *209*, 012019. https://doi.org/10.1088/1757-899X/209/1/012019

[48] Correa, D.R.N.; Vicente, F.B.; Oliveira, A.R.; Lourenc, M.L.; Kuroda, P.A.B.; Buzala, M.A.R.; Grandini, C.R. Effect of the substitutional elements on the microstructure of the Ti-15Mo-Zr and Ti-15Zr-Mo systems alloys, *J. Mater Res Technol* **2015**, *4(2)*, 180-185. https://doi.org/10.1016/j.jmrt.2015.02.007

[49] Donachi, M.J. Titanium: A Technical Guide, *ASM INTERNATIONAL*, **1988**.

[50] Minciună, M.G. Contribuții privind îmbunătățirea proprietăților aliajelor de cobalt utilizate în aplicații medicale-Teză de doctorat, Iași, **2014**.

[51] Engelbrecht, M.E.; Treurnicht, N.F.; Akdogan, G.; Sacks, N. Functional performance and machinability of titanium alloys for medical implants: A review. *In Proceedings of the SAIIE25*, Stellenbosch, South Africa, **2013**, 1–14.

[52] Al-Mobarak, N.A.; Al-Swayih, A.A.; Al-Rashoud, F.A. Corrosion behavior of Ti-6Al-7Nb alloy in biological solution for dentistry applications. *Int. J. Electrochem. Sci.* **2011**, *6*, 2031–2042.

[53] Oshida, Y.; Tuna, E.B.; Aktoren, O.; Gencay, K. Dental implant systems. *Int. J. Mol. Sci.* **2010**, *11*, 1580–1678. https://doi.org/10.3390/ijms11041580

[54] Oliveira, N.T.C.; Guastaldi, A.C. Electrochemical stability and corrosion resistance of Ti–Mo alloys for biomedical applications. *Acta Biomaterialia* **2009**, *5*, 399–405. https://doi.org/10.1016/j.actbio.2008.07.010

[55] Jeong, Y.H.; Choe, H.C.; Brantley, W.A. Nanostructured thin film formation on femtosecond laser-textured Ti-35Nb-xZr alloy for biomedical applications, *Thin Solid Films* **2011**, *519(15)*, 4668–4675. https://doi.org/10.1016/j.tsf.2011.01.014

[56] Song, Y.H.; Kim, M.K.; Park, E.J.; Song, H.J.; Anusavice, K.J.; Park, Y.J. Cytotoxicity of alloying elements and experimental titanium alloys by WST-1 and agar overlay tests *Dent. Mater.* **2014**, *30(9)*, 977–983. https://doi.org/10.1016/j.dental.2014.05.012

[57] Kolk, A.; Handschel, J.; Drescher, W.; Rothamel, D.; Kloss, F.; Blessmann, M.; Heiland, M.; Wolff, K.D.; Smeets, R. Current trends and future perspectives of bone substitute materials - from space holders to innovative biomaterials. *J. Craniomaxillofac. Surg.* **2012**, *40*, 706–718. https://doi.org/10.1016/j.jcms.2012.01.002

[58] Liu, X.; Chu, P.; Ding, C. Surface modification of titanium, titanium alloys, and related materials for biomedical applications. *Mater. Sci. Eng. R Rep.* **2004**, *47*, 49–121. https://doi.org/10.1016/j.mser.2004.11.001

[59]   Jeong, Y.-H.; Choe, H.-C.; Brantley, W.A. Silicon-substituted hydroxyapatite coating with Si content on the nanotube-formed Ti–Nb–Zr alloy using electron beam-physical vapor deposition. *Thin Solid Films* **2013**, *546*,   189–195. https://doi.org/10.1016/j.tsf.2013.05.130

[60]   Mahltig, B.; Soltmann, U.; Haase, H. Modification of algae with zinc, copper and silver ions for usage as natural composite for antibacterial applications. *Mater. Sci. Eng. C Mater. Biol. Appl.* **2013**, *33*, 979–983. https://doi.org/10.1016/j.msec.2012.11.033

[61]   Ferraris, S.; Spriano, S. Antibacterial titanium surfaces for medical implants. *Mater. Sci. Eng. C Mater. Biol. Appl.* **2016**, *61*, 965–978. https://doi.org/10.1016/j.msec.2015.12.062

[62]   Lan, M.Y.; Liu, C.P.; Huang, H.H.; Lee, S.W. Both enhanced biocompatibility and antibacterial activity in Ag-decorated TiO_2 nanotubes. *PLoS ONE* **2013**, *8*, e75364. https://doi.org/10.1371/journal.pone.0075364

[63]   Saravanan, S.; Nethala, S.; Pattnaik, S.; Tripathi, A.; Moorthi, A.; Selvamurugan, N. Preparation, characterization and antimicrobial activity of a bio-composite scaffold containing chitosan/nanohydroxyapatite/nano-silver for bone tissue engineering. *Int. J. Biol. Macromol.* **2011**, *49*, 188–193. https://doi.org/10.1016/j.ijbiomac.2011.04.010

[64]   Bauer, S.; Schmuki, P.; von der Mark, K.; Park, J. Engineering biocompatible implant surfaces. *Prog. Mater. Sci.* **2013**, *58*, 261–326. https://doi.org/10.1016/j.pmatsci.2012.09.001

[65]   Bălțatu, I.; Vizureanu, P.; Ciolacu, F.; Achiței, D.C.; Bălțatu, M.S.; Vlad, D. In Vitro study for new Ti-Mo-Zr-Ta alloys for medical use, *IOP Conf. Ser.: Mater. Sci. Eng.* **2019**, *572*, 012030. https://doi.org/10.1088/1757-899X/572/1/012030

[66]   Cumpătă, C.N. Implaturi acoperite chimic cu hidroxiapatita biologică, *Ed. Printech*, București **2012**.

[67]   Linkow, L.I. The blade vent-a new dimension in endosseous implantology, *Dent Concepts*, **1968**, *11(2)*, 3-12.

[68]   Linkow, L.I. Prefabicated mandibular prostheses for intraosseous implants, *J Prosthet Dent.* **1968**, *20(4)*, 367–375. https://doi.org/10.1016/0022-3913(68)90234-5

[69]   Antoniac, I. Biomateriale metalice utilizate la executia componentelor endoprotezelor totale de sold, *Ed. Printech*, București, **2007**.

[70]   Baltatu, M.S.; Vizureanu, P.; Balan, T.; Lohan, M.; Tugui, C.A. Preliminary Tests for Ti-Mo-Zr-Ta Alloys as Potential Biomaterials, *Book Series: IOP Conference*

Series-Materials Science and Engineering **2018**, *374*, 012023.
https://doi.org/10.1088/1757-899X/374/1/012023

[71]   Bălţatu, M.S.; Vizureanu, P.; Goanţă, V.; Tugui, C.A. C.A.; Voiculescu, I.
Mechanical tests for Ti-based alloys as new medical materials, *IOP Conf. Ser.: Mater.
Sci. Eng.*, **2019**, *572*, 012029. https://doi.org/10.1088/1757-899X/572/1/012029

[72]   C. Demian, Cercetări privind comportarea materialelor destinate implantării
osoase conform normelor europene de calitate - teză de doctorat, Universitatea
"Politehnica", Timişoara, **2007**.

[73]   H. Vermeşan, Cercetări privind comportarea la coroziune a otelurilor inoxidabile
supuse deformarii plastice si nitrurarii ionice - teza de doctorat, Universitatea Tehnică
din Cluj Napoca, **1998**.

[74]   N. Dumitraşcu, Biomateriale şi biocompatibilitate, Ed. Universităţii "Alexandru
Ioan Cuza", Iaşi, **2007**.

[75]   Tarniţă, D; Tarniţă, DN; Bîzdoacă, N; Mîndrilă, I; Vasilescu, M. Properties and
medical applications of shape memory alloys. *Rom J Morphol Embryol* **2009**, *50(1)*,
15-21.

[76]   Hunter, T.B.; Yoshino, M.T.; Dzioba, R.B.; Light, R.A.; Berger, W.G. Medical
Devices of the Head, Neck, and Spine. *RadioGraphics* **2004**, *24,* 257–285.
https://doi.org/10.1148/rg.241035185

[77]   Stuermer, E.K.; Sehmisch, S; Frosch, K.H.; Rack, T.; Dumont, C.; Tezval, M.;
Stuermer, K.M. The Elastic Bridge Plating of the Forearm Fracture: A Prospective
Study. *European Journal of Trauma and Emergency Surgery* **2009**, *35*, 147–152.
https://doi.org/10.1007/s00068-008-8002-3

[78]   Information on https://orthopedicimplantsindia.wordpress.com/2016/04/29/types-
of-orthopedic-bone-screws/. Available online: (accessed on September 10, 2021).

[79]   Bjursten, L.M.; Rasmusson, L.; Oh, S.; Smith, G.C.; Brammer, K.S.; Jin, S.
Titanium dioxide nanotubes enhance bone bonding in vivo. *J. Biomed. Mater. Res. A*
2010, *92*, 1218–1224. https://doi.org/10.1002/jbm.a.32463

[80]   Sul, Y. The significance of the surface properties of oxidized titanium to the bone
response: Special emphasis on potential biochemical bonding of oxidized titanium
implant. *Biomaterials 2003*, **24**, 3893–3907. https://doi.org/10.1016/S0142-
9612(03)00261-8

[81]   Mishra, S.K.; Teotia, A.K.; Kumar, A.; Kannan, S. Mechanically tuned
nanocomposite coating on titanium metal with integrated properties of biofilm

inhibition, cell proliferation, and sustained drug delivery. *Nanomed. Nanotechnol. Biol. Med.* **2017**, *13*, 23–35. https://doi.org/10.1016/j.nano.2016.08.010

[82] Williams, R.; Mihok, P.; Murray, J. Novel antibiotic delivery and novel antimicrobials in prosthetic joint infection. *J. Trauma Orthop.* **2016**, *4*, 52–54.

[83] Amorosa, L.F.; Jayaram, P.R.; Wellman, D.S. et al. The use of the 95-degree-angled blade plate in femoral nonunion surgery. *Eur J Orthop Surg Traumatol* **2014**, *24*, 953–960. https://doi.org/10.1007/s00590-013-1267-1

[84] Cionca, D.; Georgescu, N. Osteosinteza fracturilor deplasate de col femural la vârstnici - o alternativă terapeutică viabilă. *Jurnalul de Chirurgie-Iaşi* **2007**, 3(4).

[85] Information on https://surgeryreference.aofoundation.org/orthopedic-trauma/pediatric-trauma/distal-femur/33-m-32/external-fixation. Available online: (accessed on September 10, 2021).

CHAPTER 3

Cobalt Alloys

Silver metal cobalt is delicate, has a high liquefying point and is esteemed for its wear opposition and the capacity to hold its obstruction at high temperatures.

It is one of three normal attractive metals (iron and nickel being the other two) and holds its attraction at a higher temperature (2012°F, 1100°C) than some other metal. All in all, the cobalt has the most elevated Curie point, all things considered. Cobalt additionally has significant synergist properties [1].

The word cobalt traces all the way back to the German expression kobold of the XVI century, which means trolls or underhanded soul. Kobold was utilized in the depiction of cobalt metals that, while softened for their silver substance, set trioxide free from toxic arsenic.

The most punctual use of cobalt was in compounds utilized for blue colors in ceramics, glass and frosts. Egyptian and Babylonian earthenware painted with cobalt mixtures can be dated from 1450 preceding Christ.

In 1735, the Swedish scientific expert Georg Brandt was quick to disengage the component from copper metal. He exhibited that the blue color comes from cobalt, not from arsenic or bismuth as chemists at first suspected. After protection, cobalt metal stayed uncommon and was once in a while utilized until the 20th century.

Not long after 1900, the American business person Elwood Haynes fostered a new, consumption safe compound, which he called a starlite. Licensed in 1907, star compounds contain high substance of cobalt and chromium and are totally non-attractive [2].

One more critical advancement for cobalt accompanied the making of aluminum-nickel-cobalt magnets (AlNiCo) during the 1940s. AlNiCo magnets were the principal trade for electromagnets. In 1970, the business was additionally changed by the improvement of samarium-cobalt magnets, which gave attractive energy densities.

The modern significance of cobalt prompted the way that the London Metal Exchange (LME) presented cobalt fates in 2010.

Cobalt happens normally in nickel-bearing laterites and nickel-copper sulfide stores, and in this manner is frequently separated as a result of nickel and copper. As per the Cobalt Development Institute, around 48% of cobalt creation comes from nickel minerals, 37% from copper metals and 15% from essential cobalt creation.

The primary cobalt minerals are cobalt, erythrite, glaucodot and skutterudite. The extraction method utilized for the creation of refined cobalt metal relies upon whether the feed material is as (1) copper-cobalt sulfide mineral, (2) cobalt-nickel sulfide concentrate, (3) arsenic metal or (4) nickel-lateriteore.

After copper cathodes are created from copper sulfides containing cobalt, the cobalt, alongside different contaminations, is left on the exhausted electrolyte. Pollutions (iron, nickel, copper, zinc) are eliminated, and cobalt is hastened in its hydroxide structure utilizing lime. The metal cobalt would then be able to be refined from it utilizing electrolysis, before it is squashed and degassed to create an unadulterated, business grade metal [2].

Nickel sulfide metals containing cobalt are dealt with utilizing the Sherritt cycle, named after Sherritt Gordon Mines Ltd. (presently Sherritt International). In this cycle, the sulfide concentrate containing under 1% of cobalt is filtered under tension at high temperatures in an answer of smelling salts. Both copper and nickel are taken out in a progression of synthetic decrease processes, leaving just nickel and cobalt sulfides. Strain lysing with air, sulfuric corrosive and smelling salts recuperates more nickel before the cobalt powder is added as seeds to hasten cobalt into a vaporous hydrogen air.

Arsenide metals are simmered to eliminate most arsenic oxide. The metals are then treated with hydrochloric corrosive and chlorine, or sulfuric corrosive, to make a draining arrangement that is cleansed. From this cobalt is recuperated by electrorefining or carbonate precipitation.

Nickel-cobalt laterite metals can be softened and isolated utilizing pyrometallurgical methods or hydrometallurgical procedures, which use arrangements of sulfuric corrosive or swooning with alkali [3].

As indicated by assessments of the US Geological Survey (USGS), the worldwide cobalt mine creation was 88,000 tons in 2010. The biggest cobalt mineral delivering nations at the time were the Democratic Republic of Congo (45,000 tons), Zambia (11,000 tons) and China (6,200 tons).

Refining of cobalt frequently happens outside the nation where the mineral or cobalt concentrate is initially delivered. In 2010, the nations that created the biggest amounts of refined cobalt were China (33,000 tons), Finland (9,300 tons) and Zambia (5,000 tons). The biggest producers of refined cobalt incorporate OM Group, Sherritt International, Xstrata Nickel and Jinchuan Group [3].

Superalloys, for example, stelites, are the biggest purchaser of cobalt metal, representing around 20% of interest. Made prevalently of iron, cobalt and nickel, however containing more modest measures of different metals, including chromium, tungsten, aluminum and titanium, these superior combinations are impervious to high temperatures, consumption and wear and are utilized for the production of turbine cutting edges for fly motors, hard-confronted machine parts, exhaust valves and weapon barrels [4,5].

One more significant use for cobalt is in wear-safe combinations (for instance, Vitallium), which can be found in muscular and dental inserts, just as prosthetic hips and knees.

Hard metals, in which cobalt is utilized as a limiting material, devour around 12% of the absolute cobalt. These incorporate established carbides and jewel instruments that are utilized in cutting applications and mining apparatuses [6].

Cobalt is additionally used to deliver super durable magnets, for example, the previously mentioned AlNiCo magnets and samarium-cobalt magnets. Magnets represent 7% of the cobalt metal interest and are utilized in attractive recording media, electric engines, just as generators.

Regardless of the many employments of cobalt metal, the essential utilizations of cobalt are in the synthetic area, which represents about portion of the absolute worldwide interest. Synthetics in cobalt are utilized in metal cathodies of battery-powered batteries, just as in petrochemical impetuses, clay colors and glass cheap seats.

3.1. Properties

Cobalt is a reflexive, fragile metal that is utilized to create amazing, erosion and hotness safe composites, long-lasting magnets and hard metals (Figure 3.1) [7].

Properties:

- ➢ Atomic symbol: Co;
- ➢ Atomic number: 27;
- ➢ Atomic mass: 58.93g/mole;
- ➢ Item Category: Transition Metal;
- ➢ Density: 8.86 g / cm 3 to 20 ° C;
- ➢ Melting point: 2723°F (1495°C);
- ➢ Boiling point: 2927°C (5301°F);
- ➢ Hardness of Mohs: 5.

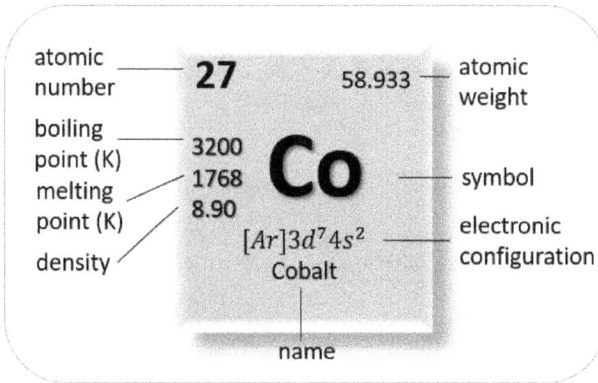

Figure 3.1. Chemical element Cobalt [8].

Composition and physico-mechanical qualities [9]:

a. Cobalt 65-67% - is a silver-like metal gives the alloy hardness.

b. Chromium 15-30% - it is white-silver metal gives hardness and high anticorrosive qualities, due to the formation of a protective film of chrome oxides at the surface of the alloy.

c. Mo(molybdenum) 5-18% - white-silver metal, malleable gives the alloy microcrystalline (micro granulation) structure, thus increasing the hardness and tensile strength.

d. Ni(nickel) 3-18% - is a white-gray, non-malleable metal that improves structural homogenization, increases plasticity, combats oxidation.

is. Mn(manganese) 0.3-2% - improves the fluidity and qualities of alloy casting, lowers the melting temperature and contributes to the removal of sulfurous gases and bonds from the alloy.

f. C (carbon) 0.2-0.4% - lowers the melting temperature and improving the fluidity of the alloy.

g. Si (silicon) 0.5% - improves fluidity and casting qualities.

h. N(nitrogen) 0.1% - contributes to the decrease of the melting temperature, improves the fluidity and also increasing the amount of N to 1% can worsen the plasticity of the alloy.

i. small amounts of Mg, Be, Ti, W (tungsten).

These composites are for the most part called cobalt-chromium combinations. On a basic level, there are two kinds of such amalgams; one is CoCrMo composite, which is for the most part utilized for projecting an item, and the other is CoNiCrMo combination, which is typically handled by hot producing. CoCrMo amalgam has been utilized for a long time in dentistry and in the assembling of fake joints. CoNiCrMo fashioned combination has been utilized in the production of the closures of prostheses, particularly on account of joints on which a high-pressure act, like the knee and hip [10].

Table 3.1. The chemical composition of Co-based alloys [11].

Element	Co28Cr6Mo (F75) poured		Co20Cr15W10Ni (F90) Wrought		Co28Cr6Mo (F1537) Wrought		Co35Ni20Cr10Mo (F562)	
	Min.	Max.	Min.	Max.	Min.	Max.	Min.	Max.
Cr	27.0	30.0	19.0	21.00	26.0	30.0	19.0	21.0
Mb	5.0	7.0	-	-	5.0	7.0	9.0	10.5
Ni	-	2.5	9.0	11.0	-	1.0	33.0	37.0
Fe	-	0.75	-	3.0	-	0.75	9.0	10.5
C	-	0.35	0.05	0.15	-	0.35	-	0.025
And	-	1.00	-	1.00	-	1.0	-	0.15
Mn	-	1.00	-	2.00	-	1.0	-	0.15
W	-	0.20	14.0	16.0	-	-		
P	-	0.020	-	0.040	-	-	-	0.015
S	-	0.010	-	0.030	-	-	-	0.010
N	-	0.25	-	-	-	0.25	-	-
Of	-	0.30	-	-	-	-	-	-
Bo	-	0.01	-	-	-	-	-	0.015
Your							-	1.0
Co.				Balance				

The American norm (ASTM) depicts 4 sorts of such combinations suggested in careful inserts: (1) Cast CoCrMo compound (F76), (2) fashioned CoCrWNi composite (F90), (3) CoNiCrMo manufactured (F562) and CoNiCrMoWFe amalgam (F563) produced. The synthetic arrangements are momentarily displayed in table 3.1 except for the F563 compound [10].

Currently, only two of the four alloys are used a lot in the manufacture of implants, namely CoCrMo alloy for casting and CoNiCrMo alloy for forging. As can be seen in table 3.1, the composition of alloys differs.

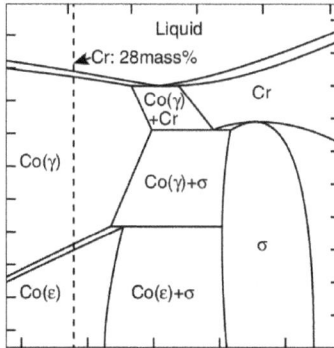

Figure 3.2. Diagram of Co-Cr phases [12,13].

The two fundamental components of the Co-based combination structure a strong arrangement containing up to 65 % by weight of Co, and the rest Cr, as shown by figure 3.2. Molybdenum is added to acquire better defects, which lead to expanded opposition following fashioning or projecting.

One of the most promising Co-based forging alloys is the CoNiCrMo alloy, originally called the MP35N (Standard Pressed Co.), which contains about 35% Co and Ni weight respectively. The alloy exhibits an advanced degree of corrosion resistance in high water (containing chloride ions) under a very high pressure. Cold processing can considerably increase the strength of the alloy. However, cold processing is difficult, especially in the case of manufacturing large devices, such as the base of the hip implant. Only hot processing can be used in the manufacture of an implant of this alloy [14,15].

The abrasive properties of the CoNiCrMo forged alloy are similar to those of the CoCrMo casting alloy (approximately 0.14 mm/year in the hip implant simulation tests). Such a wear rate is unacceptable on an implant itself. However, the former is not recommended for the free parts of joint prostheses because of the low degree of friction towards itself or towards other materials. The maximum fatigue resistance and maximum tensile strength of the CoNiCrMo forged alloy make it suitable in high-durability implants without inducing fractures or fatigue tensions. This is the case of joint prostheses at the hip. The advantage is much appreciated and visible when one implant needs to be replaced with another, because it is very difficult to replace a failed implant, especially if it is positioned deep in the femoral medullary canal. Moreover, a restored arthroplasty is usually inferior to the initial one in terms of its functions, this being much less fixed in the implant than the previous one.

The process for the experimental determination of the amount of Ni released from CoNiCrMo alloy and 316L stainless steel at 37°C in Ringer solution produced an interesting result. Although in the case of Co-based alloy, the initial amount of Ni ions released in the solution is higher, the release rate was approximately the same (3 × 10 -10g/cm²/day) for both alloys. This is surprising because the amount of Ni in the CoNiCrMo alloy is about 3 times that of the 316L alloy [16].

The modulus of elasticity for Co-based alloys varies between 220 and 234 GPa. These values are higher than in the case of other materials, such as stainless steels. Cold processing or hot treatment procedures have a small effect on the elasticity module but have a substantial effect on strength and hardness. Different elasticity modules may have implications due to the different way of transferring pressure to the bone, although the effects of some high-value modules have not been clearly established. When it is desired to reduce the rigidity of an implant, you can use Ti alloys that have an elasticity module and a density whose value is equal to half the values corresponding to Co alloys (Table 3.2) [17].

Table 3.2. Mechanical properties of alloys based on Co [11].

State	Resistance to breaking, min, ksi, (mpa	Yield limit (0.2% balance), min, ksi (mPa)	Fatigue resistance ksi (mpa)	Elongation min, %	Area reduction min, %
Co28Cr6Mo (75) poured	95(655)	65(450)	45(310)	8	8
Co20Cr15W10Ni(F90) Hardened	125(860)	45(310)	-	30	-
Co28Cr6Mo(F1537) hardened[a]	130(897)	75(517)	-	20	20
hot-processed	145(1000)	101(700)	-	12	12
processed at medium temperature	170(1172)	120(827)	-	12	12
Co35Ni20Cr10Mo(F562) hardened[b]	115(793)	35(241)	49,3(340)	50	65
	145(1000)	65(448)			
Cold processed, aged[c]	260(1793)	230(1586)	-	8	35

3.2. Classification, influence of alloying elements on cobalt properties

Cobalt-chromium alloys

Alloys based on cobalt are the most commonly used in cast or cast and quality state. They allow the manufacture of implants with specific and complicated models. Surgical alloys type cobalt-chromium are based on a cobalt and chromium system known for its excellent corrosion resistance. In the composition enter chromium 27-30% and molybdenum 5-7%. Tungsten is added to increase ductility. As in the case of stainless steel, chromium generates a layer with a high degree of passivity, which contributes to the increase of corrosion resistance. Co-Cr-Mo alloy (ASTM F 75) has a corrosion resistance superior to ASTMF 138 austenitic stainless steel, especially to crevasse corrosion. Medical devices made of Co-Cr-Mo alloys are currently produced by using the hot isostatic pressing process that helps to obtain devices with better strengths and mechanical characteristics than those obtained by deformation in the mold [18, 19].

The main alloying elements in Cobalt-based alloys are cobalt, between 53-70% and chromium, between 20 and 30% molybdenum, up to 10% in the composition are added small amounts of silicon, manganese, iron, nickel. Carbon participates between 0.05 and 0.4%. To obtain the necessary properties add modifiers such as: Ga, Zr, B, W, Nb, Si, Mg, Ta, Ti [20].

Another group of compositions includes replacement alloys in which most of the cobalt is replaced by nickel and iron. Thus, alloys are obtained that have transition properties from cobalt-based alloys to stainless steels. These alloys are hardened mainly by the formation of carbides. Therefore, the presence of carbon is of utmost importance.

Cobalt-chromium alloys (Table 3.3) have mechanical strength, corrosion resistance and good wear resistance. The lesser cobalt-chromium PFM alloys are smaller in number. In addition to base metals, they also contain nickel, tungsten and molybdenum. Tungsten and molybdenum are high temperature hardeners.

Table 3.3. Common alloys based on cobalt [20].

Material	CoCrMo Poured	CoCrMo Wrought	CoNiCrMo Poured	CoNiCr Poured	CoNiCrMo Wrought	CoNiCr Wrought
Density (gr/cm^3)	7.8	9.15	-	-	-	-
E (GPa)	200	230	-	-	-	-
Hardness (HV)	300	240	-	-	-	-
$\sigma_{0.2\%}$ (MPa)	455	390	240-450	275	1585	825-1310
UTS σ (MPa)	655	880	795-1000	600	1795	1000-1585
Elongation (%)	10	30	50	50	8	18

Cobalt contributes to the appearance of the continuous phase that ensures the basic properties; the secondary phases based on Co, Cr, Mo, Ni and C ensure the resistance, four times higher than the compact bone, and the wear resistance of the surface; chromium ensures corrosion resistance through the oxide formed on the surface, while molybdenum ensures hardness and corrosion resistance in volume. All these elements are important, because their concentration is the one that has the greatest importance in the control of

Materials Research Forum LLC
https://doi.org/10.21741/9781644901779

melting and manufacturing technologies. Also, in these alloys are present, in lower concentrations, Ni, Mn and C. Nickel has been identified in biocorrosion products, and carbon must be strictly controlled to maintain mechanical properties, such as ductility [21].

In general, cast cobalt alloys are the least ductile alloys in metal systems used for dental surgical implants, as bending must be avoided. Due to the fact that many of the devices made from these alloys were manufactured in dental laboratories, all aspects of quality control and analysis for surgical implants must be followed during the choice of the alloy, its casting and finishing. Important aspects include chemical analysis, mechanical properties and surface finishing, as specified in ASTM F 4 for surgical implants. When manufactured correctly, the implants in this group of alloys showed excellent biocompatibility [22].

These alloys are prone to oxidation during melting, and the cast alloy is brittle and hard. The high melting temperature requires the use of a phosphate fuse model, while the cooling shrinkage of about 1.9% linear, makes it difficult to achieve the dimensional accuracy required by the model. Thus, these alloys are not recommended for precision casting, such as dental crowns or bridges, while their use as a support for molten porcelain is not recommended due to the oxidation suffered by the alloy at working temperature. However, the material has good corrosion resistance and is well tolerated in the oral cavity. The typical application of these alloys is that given by Co-Cr-Mo which is used to obtain frames for partial teeth [23].

Co-Cr alloys are metastable and crystallize in the cube system with centered faces, and the carbides they present are at the grain boundary or in the interdendritic areas. The melting temperature is in the range of 1250 - 1450° C and is above the melting capacity achieved by a flame based on natural gas; thus, it is recommended to melt them in electromagnetic induction furnaces or with the help of oxyacetylene flame. The hardening of the cast alloy will lead to the appearance of a fine precipitate of carbides, which will form inside the grains and which can cause the unwanted hardening of the material. On the other hand, the slow cooling of the cast alloy will lead to the preferential orientation of the carbides at the grain boundary, where a continuous layer will form. This last aspect results in a fragile material. The compromise between these situations is given by a gradual cooling that will result in the appearance of discontinuous carbides at the grain boundary [24,25].

The value of the modulus of elasticity for these alloys is three times higher than that of gold alloys. This increased stiffness is useful for thin sections with the same loading characteristics as gold alloys (partial tooth connectors), but unfortunately this effect is accompanied by a drastic reduction in elasticity (approximately 520 MPa). The increase of the modulus of elasticity, associated with the reduction of the elasticity limit will make the design of landmarks such as staples, very difficult. The difficulty in this case is that the arms of the clip must be loosened, and this desideratum must be elastic, which implies a tension lower than the elastic limit. Due to the increased modulus for Co-Cr alloys, a higher stress is required for a given deformation, and in many cases this stress exceeds the yield strength of the material [26].

Influence of components on the properties of the alloy based on Co-Cr

The influence of the alloying elements in the case of this alloy is very important, as follows. Thus, the chromium content is responsible for the oxidation resistance, but its value exceeding the proportion of 30%, will lead to difficulties in casting. It also forms a friable phase known as the sigma phase (σ). In general, the proportions of Co and Ni are interchangeable up to a certain value. Cobalt increases modulus of elasticity, strength and hardness more than nickel. The effect of other alloying elements is much more pronounced. The safest way to increase Co-Cr alloys is to increase their carbon content. A change of only 0.02% in these alloys changes the properties to such an extent that the alloy can no longer be used in dentistry. If the proportion of carbon increases by 0.2% above the allowable value, the alloy becomes too hard and too crumbly and can no longer be used in dental prosthetics. On the other hand, reducing the carbon content by 0.2% lowers the tensile strength and yield strength so much that the alloy will have the same fate as in the previous situation [27,28].

The presence of 3 - 6% Mo contributes to the strength of the alloy. Some alloys also contain W which, although it increases strength, reduces elongation more than molybdenum.

The presence of nitrogen can only be controlled if the casting takes place in an argon atmosphere or in a vacuum. This element if it exceeds the value of 0.1% will decrease the ductility of the cast parts.

Cobalt: determines the mechanical properties and decreases the viscosity of the liquid alloy, while ensuring chemical stability.

Molybdenum: increases ductility and chemical stability, and due to the high melting point (2622°C) ensures fine granulation of the alloy.

Manganese: a deoxidizing agent during the melting of the alloy by the formation of oxides, which migrate to the surface, forming a slag that is removed.

Silicon: influences the viscosity, allowing the casting of fine shapes.

Most alloys do not contain carbon: thus, no carbides are formed, as in the case of Co-Cr alloys for skeletal prostheses. For this reason, Co-Cr alloys for joint gnato-prosthetic devices have a lower hardness.

Molybdenum-free Co-Cr alloys contain tungsten (W) as a substitute. To ensure similar properties to the alloy, tungsten must be found in double amounts compared to molybdenum. Tungsten alloys have a higher density ($\rho = 8.7$ g/cm^3) than Co-Cr-Mo alloys ($\rho = 8.2$-8.4 g/cm^3) [29].

Elgiloy alloy

In orthodontics, the most famous Cr-Co alloy is called Elgiloy proposed by the company Rocky Mountain Orthodontics USA, used for half a century for making orthodontic threads. It is a malleable alloy at room temperature, it can be heat treated to increase its elasticity.

Composition Elgiloy: 40% Co, 20% Cr, 15% Ni, 16% Fe, 7% Mo, 2% Mn, 0.4% Be, 0.4% C.

Qualities:

- It is a rigid alloy with low stored resilience;

- Possesses good malleability;

- It is biocompatible;

- Possesses high stability.

There are 4 types of Elgiloy with increasing elasticity and resilience:

- Blue Elgiloy - the softest of them, can be bent or deformed with the hand and the spike. It was used in the Edgwise technique, for making lingual springs, retainers, it can be glued and welded, the heat treatment increases its resistance to deformation.

- Yellow Elgiloy - more elastic than blue, relatively ductile (can be doubly deformed), heat treatment improves these qualities.

- Green Elgiloy - elastic, subjected to bending and plastic deformation until its heat treatment.

- Red Elgiloy - the most elastic, has high calamities of flexibility, does not undergo heat treatment.

The advantage of CrCo springs over stainless steel ones is:

- High resistance to wear and deformation

- Allow well-controlled dental movements on the perimeter

Risks of harmful action on the body:

➢CrCo alloys are passive or inert due to the presence of the protective film of chromium oxides. Patients with nickel sensitivity tolerate them well.

➢According to the WHO, chromium is a carcinogen.

➢Clinical cases of causing allergic reactions by chromium are described in the form of: generalized eczematous dermatitis, allergic contact dermatitis, hyperemia and edema.

➢Cobalt can be a health hazard only in very high doses, when it can cause polycythemia, hyperthyroidism, cardiovascular diseases [30-32].

Casting Alloys Ni-Cr

Ni-Cr compounds went to the consideration of analysts with the restrictions found in Co-Cr combinations, specifically low malleability, high shrinkage on hardening and a high inclination to oxidize. In the business these composites are known as NIMONIC and have applications in fly motor innovation. The overall substance structure of these composites is:

➢Ni: 68-80%;

➢Cr: 10-25%;

➢Mo: 0-13% - increases corrosion resistance;

➢W: 0-7% - increases the coefficient of thermal expansion;

➢Mn: 0-6%;

➢Be: 0-2% - reduces the melting temperature, increases ductility;

➢C: 0.1-0.2%.

   * (all percentages were expressed in mass percentages)

Along with these elements are also found in lower concentrations Al, Ti, Co (hardening elements) and B, Si (melt deoxidation elements). These alloys are excellent as a support for molten porcelain. The alloys crystallize in the cube system with centered faces, and the cast material has a structure with large grains, which indicates a dendritic structure. Traditionally, these materials have a higher ductility than Co-Cr alloys, but in turn vary depending on the chemical composition and the heat treatment applied. The hardening mechanism involves the precipitation of the subsequent phase or phases, referred to in particular as the "γ-phase", consisting of $(NiCo)_3(AlTi)$. Carbides can form interdendritically. The modulus values of elasticity and hardness are slightly lower than those presented by Co-Cr. The shrinkage on solidification is 1.5%, and the alloys normally melt in induction furnaces and are poured into phosphate forms. Due to the low temperature range, Ni-Cr alloys offer a much more precise casting, which makes dental bridges and crowns have minimal deviations [33,34].

3.3. Applications in medicine

Metals and alloys have been used in the oral cavity since antiquity, gold and its alloys being the preferred materials due to their corrosion resistance, biocompatibility and corresponding distinction.

In line with the evolution of metals in dentistry in the Middle Ages, the preferred metal remained gold, which being very soft was alloyed with Ag and Cu, later with Pt and platinum subgroup metals. Alloy gold was drawn into foils that fitted by stamping and plating [35,36].

The 20th century brings advances in the laboratory technology of dental prosthetics, as smelting / casting is promoted and expanded as a fundamental method in the processing of metals and alloys. In 1929-1930 a Cr-Co alloy called Vitallium appeared in the USA, which was designed to replace the platinum Au alloy in making the skeletal prosthesis.

The history of dental alloys has been influenced by three major factors: the modification and improvement of dental prosthetic technology, advances in metallurgy and the rise in the price of noble metals [37,38].

In recent decades, the dental materials market is experiencing a real explosion in terms of diversification of dental alloys. Thus, a rigorous assessment of the quality of these alloys is required, because biocompatibility plays an essential role.

Frequently, the choice of a dental alloy is made on the basis of experience and, if it has given good results over time, then it is considered satisfactory [39].

Recently, new alloys have emerged whose publicity and price make it difficult for the practitioner to choose the best alloy from a functional, biomechanical and biocompatible point of view.

Thus, out of the desire to use the best performing alloys in dental practice, it is absolutely necessary the close cooperation between the fields of research, processing, testing and standardization of dental metals and alloys [40].

The complete removal of metals / alloys from dentistry and the promotion of completely non-metallic alternatives - all-ceramic systems is an alternative, but with limited perspectives in our country.

Until the final elimination of metals and dental alloys, the factors that influence their structure and, implicitly, their properties in the processes of processing and dental use must be taken into account.

Due to its special anticorrosive qualities and good fluidity, they have been used in OMF for osteosynthesis [41].

And in orthopedic dentistry for making cast metal carcasses, skeletal prostheses, one-piece bridges as well as for immobilizing teeth in the form of splints.

In orthodontics, Cr-Co is widely used to make orthodontic wires and springs, retainers (similar to a splint).

Cr-Co alloy is one of the groups of stainless alloys, known for its high physical and mechanical qualities being extra-hard and relatively cheap.

On the chromatic exterior, it resembles stainless steel, but it does not fall into their category because it does not contain Fe (iron) or it is in small quantities up to 1.1%.

Cr-Co alloys feature complex combinations superior to Wiple. Currently, over 100 varieties of these alloys are used and new ones with improved qualities are being developed.

The Cr-Co alloy containing W and Mo bears the trade name Stelites, proposed by the Canadian company Deloron company [42].

Technological characteristics of Cr-Co alloy:

✓ Possesses a low density of 8 g/cm^2;

✓ It has a melting temperature between 1300 and 1500°C, which requires special packaging masses with a high melting point;

✓ Breaking strength 89.6 kg/mm^2;

✓ They are hard (extra hard), resistant to abrasion and distortion, which is why they are difficult to sand (electropolishing);

✓ Hardness after Brinell 370 kg/mm^2;

✓ Possesses low viscosity, good fluidity and pours accurately, obtaining exact, thin castings with low elasticity;

✓ Crystallizes homogeneously, with austenitic structure on cooling;

✓ Cr-Co wire has good elasticity, lower than Wipla wire;

✓ Creates perfect shine, resistant in the oral cavity [43,44].

Trade names:

- ➤ Vitalium: V260 and V180;
- ➤ Wiptam: (crochet hooks);
- ➤ Remanium;
- ➤ Elgiloy;
- ➤ Robonite;
- ➤ Nitrogen;
- ➤ Wisil.

Indications for use [45,46]:

- In orthodontics: orthodontic wires and springs, cast metal housings for fixed orthodontic appliances with high wear resistance.

- In dental prosthetics: cast metal housings for skeletal prostheses, for dental bridges.

- Manufacture of surgical instruments (Cr-Co alloy with Mo).

- Osteonsynthesis plates: due to the special anticorrosive qualities and a good fluidity.

In these times it is possible to offer a combined restoration of fixed and mobile prosthesis using only an invaluable Co-Cr dental alloy.

This alloy allows to achieve clinical performance and then demonstrate corrosion resistance. In most recent reports in the field of materials, total ceramic systems have been cited since 1980. They are still widely used for the infrastructure of ceramic restorations, especially in fixed prostheses.

Since history, precious alloys have been the most commonly used but the popularity of the base metal alloy has increased since 1970. The increased base metal alloy has demonstrated in most clinical cases good clinical performance and resistance to permanent intraoral deformities.

The highest elasticity and hardness of the base metal alloy is suitable for metal-ceramic restorations or partially-removable prostheses. The mechanical properties of this alloy and its low cost are the reasons why we consider this alloy as a good option. The potential biological risk and its difficult characteristics are the first disadvantages of this alloy [47].

Clinicians have proposed the use of non-precious alloys Ni, Cr, Mo, CoCrMo or CoCrW. Despite the economic causes, their use has increased especially in mobile prostheses and to a lesser degree in fixed prostheses.

Even if they are made thin, prosthetic constructions are 40%, 50% stiffer than precious alloys and are much lighter than the latter (about $8g/cm^3$ compared to $15g/cm^3$ for precious alloys). The presence of chromium (about 20%), tungsten (approximately 5%) and molybdenum ensure biocompatibility and increased corrosion resistance [48].

In previous years, CoCr alloys were used for partially removable works.

Currently, they are more used than NiCr alloys even for fixed prostheses. CoCr alloys contain cobalt predominantly and sometimes in small quantities and tungsten and have increased rigidity and hardness.

Electrochemical studies have shown that CoCr alloys are more resistant to corrosion than NiCr alloys. Ni-based alloys have a higher allergic potential than CoCr alloys are rarely allergic. Thus, the use of CoCr alloys, while CoCr alloys are rarely allergic [49].

Thus, the use of CoCr alloys has become a routine procedure in dental laboratories.

The biocompatibility and mechanical properties of an alloy used in dentures depend on the material and the correct manufacture of it. CoCr alloy for the components of a complete restoration can be advantageous. The disadvantages of such a choice are represented by laborious laboratory procedures, such as additional gain and good metallurgical knowledge of the need to use a single alloy.

Construction of implants using Co - based alloys is cast by an old method of waxing (injection molding) which involves the following steps [50-53]:

1. Make a mold / wax mold of the desired part.

2. The (casting) mold is coated with a refractory substance, first by coating it in a thin layer with a paste / ceramic (silicon suspension in ethyl silicate solution), followed by complete coating after drying.

3. The wax is melted in an oven (100-150°C).

4. The mold is heated to high temperature, burning any traces of wax or gas-releasing substances.

5. The molten alloy is cast by gravitational or centrifugal force. The casting temperature is about 800-1000°C, and the alloy is at 1350-1400°C [4].

Dentistry

Cobalt-based alloys have multiple uses in dentistry due to their very good properties (Table 3.4).

Materials Research Forum LLC

https://doi.org/10.21741/9781644901779

Table 3.4. Characteristics of cobalt-based alloys [54].

Specification	Cobalt-based alloys
International standards	ASTM F-75 ASTM F-799 ASTM F-1537(Cast and Forged)
The main alloying elements in %	Cr = 19-30 Ni = 0-37 Mo = 0-10
Advantages	-resistance to wear; -corrosion resistance; -resistance to fatigue.
Disadvantages	-high modules; -modest biocompatibility.
Uses	-dentistry; -rods in prosthesis.

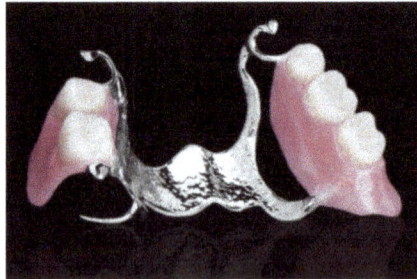

Figure 3.3. Vitallium 2000 - Partial skeletal prosthesis, dental crown [55].

The wear resistance determined by the stress on the crushing contact in mastication is an important characteristic of the metal alloys used in the fixation and partial realization of the teeth (Figure 3.3).

Metal internal fasteners

Surgical techniques use various fracture fixation devices in the form of: wire, needles, stem, screws, fracture fixing plates, intramedullary devices.

Aneurysmal clips

High purity CoNiCrMo laminated alloys (F562 - according to ASTM) are used, which have the advantage that they are more chemically stable than stainless steel and can be hardened to ensure the required elasticity.

Intervertebral disc prosthesis

The CHARITE III prosthesis consists of two concave plates made of Co-Cr-Mo alloy, hot stamped from laminated sheet, the plates have on the outer surface fixing spurs in the adjacent bone tissue. Between the metal plates was a high-density polyethylene disc, which is perfectly fitted with the concave inner space of the metal plates (Figure 3.4).

Figure 3.4. Charite III prosthesis overview [56].

Orthopedic joint prostheses

Hip joint prosthesis, an overview of the way of fixing, through orthopedic surgery of the hip prosthesis is shown in figure 3.5.

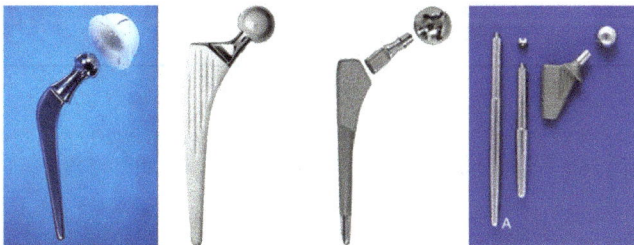

Monoblock   Uni-modular   Bi-modular   Tri-modular
Figure 3.5. Types of monobloc and modular prostheses [57].

Knee joint prosthesis

An overview of the knee prosthesis is shown in figure 3.6, where it is found that the joint consists of two main elements:

- the upper component that attaches to the lower end of the femoral bone;
- the lower component fixed on the upper end of the tibial bone.

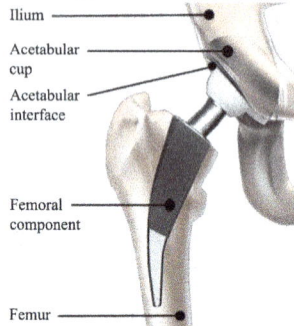

Figure 3.6. General view of the knee joint prosthesis [58].

Ankle joint prosthesis

The materials used to make ankle prostheses are Co-Cr alloys in combination with high density polyethylene (UHMWPE). From the point of view of the functional conception, the ankle joint prostheses are of two types: congruent and incongruent (Figure 3.7).

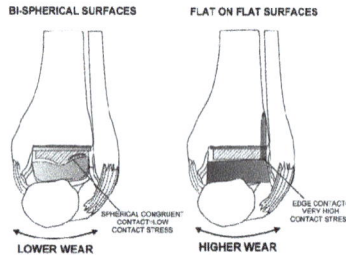

Figure 3.7. Comparison of congruent deep-sulcus type and incongruent flat-on-flat type total ankle replacements during inversion-eversion [59].

Shoulder joint prosthesis

The shoulder joint prostheses are made on the principle of the spherical joint (spherical head-cup) similar to the hip joint, as shown schematically in figure 3.8.

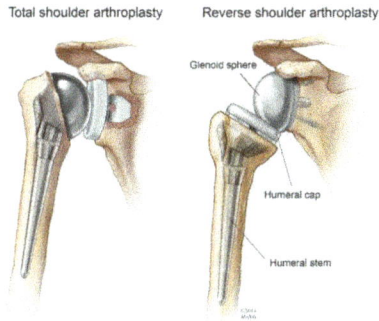

Figure 3.8. Types of shoulder joint prostheses [60].

The metallic materials used in the processing of the prosthetic joint are Co-Cr alloys joined by a cup made of high-density polyethylene, or a metal cup on which a film of carbon material (graphite) has been deposited [61].

Cardiovascular prostheses

The metals used in heart valve prostheses are the rigid support of the artificial ring valve. Also, the valve disc is made of metallic materials. The most used metals are cobalt-based alloys (Co-Cr-Mo) [62].

Co-Cr-Ni-Mo (MP35N) alloys are characterized as materials with high mechanical strength and very good corrosion resistance. MP35N alloys are indicated for the manufacture of neuromuscular stimulation electrodes and as electrodes in pacemakers [63, 64].

References

[1] Bombac, D.M.; Brojan, M.; Fajfar, P.; Kosel, F.; Turk, R. Review of materials in medical applications, *Materials and Geoenvironment* **2007**, *54(4)*, 471-499.

[2] Singh, R.; Singh, S.; Hashmi, M.S.J. Implant Materials and Their Processing Technologies; *Elsevier Ltd.: Amsterdam*, The Netherlands, **2016**; ISBN 9780128035818. https://doi.org/10.1016/B978-0-12-803581-8.04156-4

[3] Bedolla-Gil, Y.; Hernandez-Rodriguez, M.A.L. Tribological Behavior of a Heat-Treated Cobalt-Based Alloy. *J. Mater. Eng. Perform.* **2013**, *22*, 541–547. https://doi.org/10.1007/s11665-012-0261-9

[4] Øilo, M.; Nesse, H.; Lundberg, O.J.; Gjerdet, N.R. Mechanical properties of cobalt-chromium 3-unit fi xed dental prostheses fabricated by casting, milling, and additive manufacturing. *J. Prosthet. Dent.* **2015**, *120*, 1–7. https://doi.org/10.1016/j.prosdent.2017.12.007

[5] Song, C.; Yang, Y.; Wang, Y.; Wang, D.; Yu, J. Research on rapid manufacturing of Co-Cr-Mo alloy femoral component based on selective laser melting. *Int. J. Adv. Manuf. Technol.* **2014**, *75*, 445–453. https://doi.org/10.1007/s00170-014-6150-7

[6] Harun, W.; Sharuzi, W.; Kadirgama, K.; Samykano, M.; Ramasamy, D.; Ahmad, I.; Moradi, M. Mechanical behavior of selective laser melting-produced metallic biomaterials. In Mechanical Behavior of Biomaterials; Davim, J.P., *Ed.; Woodhead Publishing: Cambridge*, UK **2019**, 101–116. ISBN 9780081021743. https://doi.org/10.1016/B978-0-08-102174-3.00005-X

[7] Manea, A.; Bran, S.; Baciut, M.; Armencea, G.; Pop, D.; Berce, P.; Vodnar, D.-C.; Hedesiu, M.; Dinu, C.; Petrutiu, A.; et al. Sterilization protocol for porous dental implants made by Selective Laser Melting. *Dent. Med.* **2018**, *91*, 452–457. https://doi.org/10.15386/cjmed-987

[8] Information on https://www.britannica.com/science/cobalt-chemical-element. Available online: (accessed on September 10, 2021).

[9] Guoqing, Z.; Junxin, L.; Xiaoyu, Z.; Jin, L.; Anmin, W. Effect of Heat Treatment on the Properties of Co-Cr-Mo Alloy Manufactured by Selective Laser Melting. *J. Mater. Eng. Perform.* **2018**, *27*, 2281–2287. https://doi.org/10.1007/s11665-018-3351-5

[10]   Ramírez-Vidaurri, L.E.; Castro-Román, M.; Herrera-Trejo, M.; García-López, C.V.; Almanza-Casas, E. Cooling rate and carbon content effect on the fraction of secondary phases precipitate in as-cast microstructure of ASTM F75 alloy. *J. Mater. Process. Technol.* **2009**, *209*, 1681–1687. https://doi.org/10.1016/j.jmatprotec.2008.04.039

[11]   Bhaduri, A. Creep and Stress Repture. Mechanical Properties and Working of Metals and Alloys; *Springer: Singapore*, **2018**, 257–314. https://doi.org/10.1007/978-981-10-7209-3_7

[12]   Ishida, B.K.; Nlshizawa, T. The Co-Cr (Cobalt-Chromium) System. Bull. *Alloy Phase Diagrams* **1990**, *11*, 357–370. https://doi.org/10.1007/BF02843315

[13]   Narushima, T.; Ueda, K.A. Co-Cr Alloys as Effective Metallic Biomaterials. In: Niinomi, M.; Narushima, T.; Nakai, M. (eds) Advances in Metallic Biomaterials.

Springer Series in Biomaterials Science and Engineering **2015**, *3*. Springer, Berlin, Heidelberg. https://doi.org/10.1007/978-3-662-46836-4_7

[14]   Tonelli, L.; Fortunato, A.; Ceschini, L. Co-Cr alloy processed by Selective Laser Melting (SLM): Effect of Laser Energy Density on microstructure, surface morphology, and hardness. *J. Manuf. Process.* **2020**, *52*, 106–119. https://doi.org/10.1016/j.jmapro.2020.01.052

[15]   Kajima, Y.; Takaichi, A.; Kittikundecha, N.; Nakamoto, T.; Kimura, T.; Nomura, N.; Kawasaki, A.; Hanawa, T.; Takahash, H.; Wakabayashi, N. Effect of heat-treatment temperature on microstructures and mechanical properties of Co-Cr-Mo alloys fabricated by selective laser melting. *Mater. Sci. Eng. A* **2018**, *726*, 21–31. https://doi.org/10.1016/j.msea.2018.04.048

[16]   Yan, X.; Lin, H.; Wu, Y.; Bai, W. Effect of two heat treatments on mechanical properties of selective-laser-melted Co-Cr metal-ceramic alloys for application in thin removable partial dentures. *J. Prosthet. Dent.* **2018**, *119*, 1028.e1–1028.e6. https://doi.org/10.1016/j.prosdent.2018.04.002

[17]   Seki, E.; Kajima, Y.; Takaichi, A.; Kittikundecha, N.; Htoot, H.; Cho, W.; Linn, H.; Doi, H.; Hanawa, T.; Wakabayashi, N. Effect of heat treatment on the microstructure and fatigue strength of Co-Cr-Mo alloys fabricated by selective laser melting. *Mater. Lett.* **2019**, *245*, 53–56. https://doi.org/10.1016/j.matlet.2019.02.085

[18]   Doherty, R.D.; Hughes, D.A.; Humphreys, F.J.; Jonas, J.J.; Juul Jensen, D.; Kassner, M.E.; King, W.E.; McNelley, T.R.; McQueen, H.J.; Rollett, A.D. Current issues in recrystallization: A review. *Mater. Sci. Eng. A* **1997**, *238*, 219–274. https://doi.org/10.1016/S0921-5093(97)00424-3

[19]   Mróz, A.; Jakubowicz, J.; Gierzy´nska-dolna, M.; Wi´sniewski, T.; Wendland, J. Wpływ technologii wytwarzania wyrobów ze stopu Co28Cr6Mo na ich wła´sciwo´sci fizyczne, mechaniczne i odporno´s´c korozyjn ̨a. In´zynieria *Mater.* **2015**, *1*, 2–8. https://doi.org/10.15199/28.2015.1.1

[20]   Takaichi, A.; Suyalatu; Nakamoto, T.; Joko, N.; Nomura, N.; Tsutsumi, Y.; Migita, S.; Doi, H.; Kurosu, S.; Chiba, A.; et al. Microstructures and mechanical properties of Co-29Cr-6Mo alloy fabricated by selective laser melting process for dental applications. *J. Mech. Behav. Biomed. Mater.* **2013**, *21*, 67–76. https://doi.org/10.1016/j.jmbbm.2013.01.021

[21]   Popovich, V.A.; Borisov, E.V.; Popovich, A.A.; Su, V.S.; Masaylo, D.V.; Alzina, L. Functionally graded Inconel 718 processed by additive manufacturing:

Crystallographic texture, anisotropy of microstructure and mechanical properties. *Mater. Des.* **2016**, *114*, 441–449. https://doi.org/10.1016/j.matdes.2016.10.075

[22]   Sing, S.L.; Huang, S.; Yeong, W.Y. Effect of solution heat treatment on microstructure and mechanical properties of laser powder bed fusion produced cobalt-28chromium-6molybdenum. *Mater. Sci. Eng. A* **2020**, *769*, 138511. https://doi.org/10.1016/j.msea.2019.138511

[23]   Béreš, M.; Silva, C.C.; Sarvezuk, P.W.C.; Wu, L.; Antunes, L.H.M.; Jardini, A.L.; Feitosa, A.L.M.; Žilková, J.;de Abreu, H.F.G.; Filho, R.M. Mechanical and phase transformation behaviour of biomedical Co-Cr-Mo alloy fabricated by direct metal laser sintering. *Mater. Sci. Eng. A* **2018**, *714*, 36–42. https://doi.org/10.1016/j.msea.2017.12.087

[24]   Liu, F.; Lin, X.; Yang, G.; Song, M.; Chen, J.; Huang, W. Microstructure and residual stress of laser rapid formed Inconel 718 nickel-base superalloy. *Opt. Laser Technol.* **2011**, *43*, 208–213. https://doi.org/10.1016/j.optlastec.2010.06.015

[25]   Sun, S.H.; Koizumi, Y.; Kurosu, S.; Li, Y.P.; Matsumoto, H.; Chiba, A. Build direction dependence of microstructure and high-temperature tensile property of Co-Cr-Mo alloy fabricated by electron beam melting. *Acta Mater.* **2014**, *64*, 154–168. https://doi.org/10.1016/j.actamat.2013.10.017

[26]   Wang, J.; Ren, J.; Liu, W.; Wu, X.; Gao, M.; Bai, P. Effect of Selective Laser Melting Process Parameters on Microstructure and Properties of Co-Cr Alloy. *Materials (Basel)* **2018**, *11*, 1546. https://doi.org/10.3390/ma11091546

[27]   Kok, Y.; Tan, X.P.; Wang, P.; Nai, M.L.S.; Loh, N.H.; Liu, E.; Tor, S.B. Anisotropy and heterogeneity of microstructure and mechanical properties in metal additive manufacturing: A critical review. *Mater. Des.* **2017**, *139*, 565–586. https://doi.org/10.1016/j.matdes.2017.11.021

[28]   Averyanova, M.; Bertrand, P.; Verquin, B. Manufacture of Co-Cr dental crowns and bridges by selective laser Melting technology. *Virtual Phys. Prototyp.* **2011**, *6*, 179–185. https://doi.org/10.1080/17452759.2011.619083

[29]   Lu, Y.; Wu, S.; Gan, Y.; Li, J.; Zhao, C.; Zhuo, D.; Lin, J. Investigation on the microstructure, mechanical property and corrosion behavior of the selective laser melted Co-Cr-W alloy for dental application. *Mater. Sci. Eng. C* **2015**, *49*, 517–525. https://doi.org/10.1016/j.msec.2015.01.023

[30]   Han, X.; Sawada, T.; Schille, C.; Schweizer, E.; Scheideler, L.; Geis-Gerstorfer, J.; Rupp, F.; Spintzyk, S. Comparative analysis of mechanical properties and metal-

ceramic bond strength of Co-Cr dental alloy fabricated by different manufacturing processes. *Materials (Basel)* **2018**, *11*, 1801. https://doi.org/10.3390/ma11101801

[31]   Strub, J.R.; Rekow, E.D.; Witkowski, S. Computer-aided design and fabrication of dental restorations: Current systems and future possibilities. *J. Am. Dent. Assoc.* **2006**, *137*, 1289–1296. https://doi.org/10.14219/jada.archive.2006.0389

[32]   Javaid, M.; Haleem, A.; Kumar, L. Current status and applications of 3D scanning in dentistry. *Clin. Epidemiol. Glob. Health* **2019**, *7*, 228–233. https://doi.org/10.1016/j.cegh.2018.07.005

[33]   Dobrza´nski, L.A.; Dobrza´nski, L.B. Dentistry 4.0 concept in the design and manufacturing of prosthetic dental restorations. *Processes* **2020**, *8*, 525. https://doi.org/10.3390/pr8050525

[34]   Gu, D.; Shi, Q.; Lin, K.; Xi, L. Microstructure and performance evolution and underlying thermal mechanisms of Ni-based parts fabricated by selective laser melting. *Addit. Manuf.* **2018**, *22*, 265–278. https://doi.org/10.1016/j.addma.2018.05.019

[35]   Hassani, F.Z.; Ketabchi, M.; Bruschi, S.; Ghiotti, A. Effects of carbide precipitation on the microstructural and tribological properties of Co-Cr-Mo-C medical implants after thermal treatment. J. Mater. Sci. 2016, 51, 4495–4508. https://doi.org/10.1007/s10853-016-9762-5

[36]   Anusavice, K.; Shen, C.; Rawls, H.R. Phillips' Science of Dental Materials, 12th ed.; *Saunders: St. Louis*, MO, USA, **2012**.

[37]   Craig, R.G. Materiały Stomatologiczne, 12th ed.; *Powers*, J.M., Sakaguchi, R.L., Shaw, H., Shaw, J.G., Eds.; Edra Urban and Partner: Wrocław, Poland, **2008**; ISBN 9780323081085.

[38]   Myszka, D.; Skrodzki, M. Comparison of Dental Prostheses Cast and Sintered by SLM from Co-Cr-Mo-W Alloy. *Arch. Foundry Eng.* **2016**, *16*, 201–207. https://doi.org/10.1515/afe-2016-0110

[39]   Ferraiuoli, P.; Taylor, J.C.; Martin, E.; Fenner, J.W.; Narracott, A.J. The accuracy of 3D optical reconstruction and additive manufacturing processes in reproducing detailed subject-specific anatomy. *J. Imaging* **2017**, *3*, 45. https://doi.org/10.3390/jimaging3040045

[40]   Antanasova, M.; Kocjan, A.; Kovaˇc, J.; Žužek, B.; Jevnikar, P. Influence of thermo-mechanical cycling on porcelain bonding to cobalt–chromium and titanium

dental alloys fabricated by casting, milling, and selective laser melting. J. Prosthodont. Res. **2018**, *62*, 184–194. https://doi.org/10.1016/j.jpor.2017.08.007

[41] Reclaru, L.; Ardelean, L.C. Current Alternatives for Processing CoCr Dental Alloys Lucien; *Elsevier Inc.:Cambridge*, MA, USA, **2018**, *1–3*, ISBN 9780128051443. https://doi.org/10.1016/B978-0-12-801238-3.11100-6

[42] Haleem, A.; Javaid, M. 3D scanning applications in medical field: A literature-based review. *Clin. Epidemiol. Glob. Health* **2019**, *7*, 199–210. https://doi.org/10.1016/j.cegh.2018.05.006

[43] Singh, A.V.; Dad Ansari, M.H.; Wang, S.; Laux, P.; Luch, A.; Kumar, A.; Patil, R.; Nussberger, S. The adoption of three-dimensional additive manufacturing from biomedical material design to 3D organ printing. *Appl. Sci.* **2019**, *9*, 811. https://doi.org/10.3390/app9040811

[44] Dikova, T. Properties of Co-Cr Dental Alloys Fabricated Using Additive Technologies. In Biomaterials in Regenerative Medicine; Dobrza´nski, L.A., Ed.; *IntechOpen*: London, UK, **2018**, 141–159. https://doi.org/10.5772/intechopen.69718

[45] Oliveira, T.T.; Reis, A.C. Fabrication of dental implants by the additive manufacturing method: A systematic review. *J. Prosthet. Dent.* **2019**, *122*, 270–274. https://doi.org/10.1016/j.prosdent.2019.01.018

[46] Revilla-León, M.; Özcan, M. Additive Manufacturing Technologies Used for 3D Metal Printing in Dentistry. *Curr. Oral Health Rep.* **2017**, *4*, 201–208. https://doi.org/10.1007/s40496-017-0152-0

[47] Revilla-León, M.; Klemm, I.M.; García-Arranz, J.; Özcan, M. 3D Metal Printing– Additive Manufacturing Technologies for Frameworks of Implant- Borne Fixed Dental Prosthesis. *Eur. J. Prosthodont. Restor. Dent.* **2017**, *25*, 143–147.

[48] Gabor, A.-G.; Zaharia, C.; Stan, A.T.; Gavrilovici, A.M.; Negrut,iu, M.-L.; Sinescu, C. Digital Dentistry—Digital Impression and CAD/CAM System Applications. *J. Interdiscip. Med.* **2017**, *2*, 54–57. https://doi.org/10.1515/jim-2017-0033

[49] Okazaki, Y.; Ishino, A.; Higuchi, S. Chemical, physical, and mechanical properties and microstructures of laser-sintered Co-25Cr-5Mo-5W (SP2) and W-Free Co-28Cr-6Mo alloys for dental applications. *Materials (Basel)* **2019**, *12*, 4039. https://doi.org/10.3390/ma12244039

[50] Kajima, Y.; Takaichi, A.; Nakamoto, T.; Kimura, T.; Yogo, Y.; Ashida, M.; Doi, H.; Nomura, N.; Takahashi, H.; Hanawa, T.; et al. Fatigue strength of Co-Cr-Mo alloy

clasps prepared by selective laser melting. J. Mech. Behav. *Biomed. Mater.* **2016**, *59*, 446–458. https://doi.org/10.1016/j.jmbbm.2016.02.032

[51]   Murr, L.E.; Gaytan, S.M.; Ramirez, D.A.; Martinez, E.; Hernandez, J.; Amato, K.N.; Shindo, P.W.; Medina, F.R.; Wicker, R.B. Metal Fabrication by Additive Manufacturing Using Laser and Electron Beam Melting Technologies. *J. Mater. Sci. Technol.* **2012**, *28*, 1–14. https://doi.org/10.1016/S1005-0302(12)60016-4

[52]   Yap, C.Y.; Chua, C.K.; Dong, Z.L.; Liu, Z.H.; Zhang, D.Q.; Loh, L.E.; Sing, S.L. Review of selective laser melting: Materials and applications. *Appl. Phys. Rev.* **2015**, *2*, 1–21. https://doi.org/10.1063/1.4935926

[53]   Beaman, J.J.; Deckard, C.R. Selective Laser Sinterng with Assisted Powder Handlng. U.S. Patent 4, 938,816, **1990**.

[54]   Williams D.F., Biofunctionality and Biocompatibility, Materials Science and Technology, *Medical and Dental Materials, Weinheim* **1992**, *14*, 2-27.

[55]   Information on https://www.dentalartslab.com/products-services/removable-partial-dentures/vitallium%C2%AE-2000-plus-partial-dentures/. Available online: (accessed on September 10, 2021).

[56]   Serhan, H.; Mhatre, D.; Defossez, H.; Bono, C.M. Motion-preserving technologies for degenerative lumbar spine: The past, present, and future horizons, *SAS Journal* **2011**, *5*, 75–89. https://doi.org/10.1016/j.esas.2011.05.001

[57]   Oladokun, A. Mechanism of fretting corrosion at the modular taper interface of hip prosthesis. *PhD Thesis* **2017**.

[58]   Hafezalkotob1, A.; Hafezalkotob, A. Interval MULTIMOORA method with target values of attributes based on interval distance and preference degree: biomaterials selection. *J Ind Eng Int* **2017**, *13*, 181–198. https://doi.org/10.1007/s40092-016-0176-4

[59]   Buechel, F.F.; Pappas, M.D. Ten-Year Evaluation of Cementless Buechel-Pappas Meniscal Bearing Total Ankle Replacement. *Foot & Ankle International* **2013**, *24(6)*, 462-72. https://doi.org/10.1177/107110070302400603

[60]   Information on https://www.drleeorthopedics.com/blog/what-type-of-injuries-may-benefit-from-reverse-total-shoulder-surgery. Available online: (accessed on September 10, 2021).

[61]   Wang, W.J.; Yung, K.C.; Choy, H.S.; Xiao, T.Y.; Cai, Z.X. Effects of laser polishing on surface microstructure and corrosion resistance of additive manufactured

Co-Cr alloys. *Appl. Surf. Sci.* **2018**, *443*, 167–175.
https://doi.org/10.1016/j.apsusc.2018.02.246

[62]   Yung, K.C.; Wang, W.J.; Xiao, T.Y.; Choy, H.S.; Mo, X.Y.; Cai, Z.X.; Wang, W.J.; Xiao, T.Y.; Choy, H.S.; Mo, X.Y.; et al. Laser polishing of additive manufactured Co-Cr components for controlling their wettability characteristics. *Surf. Coat. Technol.* **2018**, *351*, 89–98. https://doi.org/10.1016/j.surfcoat.2018.07.030

[63]   Lu, Y.; Wu, S.; Gan, Y.; Zhang, S.; Guo, S.; Lin, J.; Lin, J. Microstructure, mechanical property and metal release of As-SLM Co-Cr-W alloy under different solution treatment conditions. *J. Mech. Behav. Biomed. Mater.* **2016**, *55*, 179–190. https://doi.org/10.1016/j.jmbbm.2015.10.019

[64]   Ngo, T.D.; Kashani, A.; Imbalzano, G.; Nguyen, K.T.Q.; Hui, D. Additive manufacturing (3D printing): A review of materials, methods, applications and challenges. *Compos. Part B Eng.* **2018**, *143*, 172–196. https://doi.org/10.1016/j.compositesb.2018.02.012

Advanced Metallic Biomaterials   Materials Research Forum LLC
Materials Research Foundations **118** (2022)   https://doi.org/10.21741/9781644901779

CHAPTER 4

Stainless Steel Alloys

Stainless steels are a class of metallic materials that have mostly the properties imposed on the materials used in the human body environment: biocompatibility, chemical, thermal and mechanical stability in the special conditions of the human environment [1].

Treated steels are on a very basic level iron amalgams containing essentially 10.5% chromium, albeit other alloying components (molybdenum, copper, titanium, nickel) can likewise be utilized in explicit extents to work on their construction and properties, like strength, formability and cryogenic hardness [2].

The stainless-steel work results from the chromium content to be exposed to passivation, framing an idle film of chromium oxide on a superficial level. This inactive layer forestalls further consumption by discouraging the dissemination of oxygen on the steel surface to stop the spread of erosion in the metal mass.

Stainless steels are grouped by their substance of alloyed metal, yet additionally by their translucent design. The treated steel prepares of the 300 series have an austenitic glasslike structure, which is driven on the cubic face with four iotas in the unit cell for a higher thickness. They have a most extreme carbon level of 0.15% (low carbon is essential for the properties of hardened steel), at least chromium of 16% and nickel and additionally manganese appropriate to safeguard an austenitic design at temperatures that reach from the cryogenic district to that of dissolving point of the composite [3].

However, stainless steel is not completely resistant to corrosion. Under certain conditions, such as repeated exposure to highly concentrated salt water, even stainless steel will corrode.

4.1. Properties

Chromium is a significant part of erosion safe treated steels. The base successful worth of the centralization of chromium is 11 % by weight. Chromium is a receptive component yet both it and its amalgams can be passivated to acquire fantastic consumption opposition [4].

Austenitic tempered steels, particularly types 316 and 316L, are regularly utilized in inserts. They solidify not by warming, but rather by chilly handling. The incorporation of molybdenum expands the odds of consumption opposition in salt water. ASTM (American Society of Testing Materials) recommends type 316L, not 316 in the realization of implants [5, 6].

Nickel is used in the stabilization of the austenitic phase at room temperature and, moreover, for increasing the resistance to corrosion. The stability of the austenitic phase, in the case of stainless steels with 0.10 % carbon weight, can be influenced by the content of Ni and Cr [7].

Table 4.1. Mechanical properties of stainless steels used in surgical implants [1,2].

Processing conditions	Resistance to breaking, min, ksi, (mpa)	Yield limit (0.2% balance), min, ksi (mPa)	Elongation, (%)	Hardness Rockwell Max.
		Bars and threads(F138)		
Hardening	71(490)	27,5(190)	40	-
Cold processing	125(860)	100(690)	12	-
Hardening	196(1350)	-	-	-
Cold shooting	125(860)	-	-	-
		Sheet and strip(F139)		
Hardening	71(490)	27,5(190)	40	95 HRB
Cold finished	125(860)	100(690)	10	—

Stainless steel (F138 and ASTM F139)

Table 4.1 shows the mechanical properties of hardened steels of type 316 and 316L. As can be seen, a wide assortment of properties can be acquired relying upon the most common way of warming (to get delicate materials) or cooling (for more noteworthy strength and hardness). The creator should be exceptionally cautious while picking material of this sort. Indeed, even sort 316L can erode in the human body under particular conditions, like a region with exceptionally high tension and lacking oxygen. Nonetheless, they are reasonable for use in brief embeds like bars, screws and hip joints [8,9].

Austenitic stainless steels harden very quickly following mechanical processing, so they can only be processed cold after an intermediate hot treatment. However, hot processing should not induce the formation of chromium carbide (CoCr4) at the level of flat imperfections, which can reduce Cr and C from imperfections, causing corrosion. For the same reason, austenitic stainless-steel implants are not welded [10].

The distortion of the components following hot processing can occur, but this problem can be easily solved by keeping the constant temperature under control. Another undesirable effect of hot processing is the formation on the surface of layers of oxides, which must be removed either chemically (with acids) or mechanically (by sandblasting). After removing the layers, the surface of the compound is finished until it becomes like a mirror or matte. Subsequently, the surface is cleaned, degreased, and passivated with nitric acid (ASTM F86 standard). The compound is washed and cleaned again before packaging and sterilization [11].

Corrosion resistance

The group of materials that today form the stainless-steel family has its beginning in 1913, in Sheffield, England. Trying to find new materials for the cannon pipes, Harry Brearley noticed that some of the samples he had prepared do not rust and are difficult to attack with chemicals, a stage he was following to study the microstructure under the optical microscope. All of these samples contained about 13% chromium. They were the first stainless steels. For the beginning, these steels were made of cutes, the Sheffield becoming famous for them. Today stainless steels are present all around us, the modern technique cannot be conceived without them [12].

The corrosion resistance of stainless steel comes from the surface formation of a thin protective film (Figure 4.1). It's made up of oxides that contain chromium and it's spontaneously formed in the presence of oxygen. Even if it is damaged, physically or chemically, this protective film has the property of recovering as soon as the cause that generated the damage is removed and the surface is again exposed to the action of oxygen in the air or water [13,14].

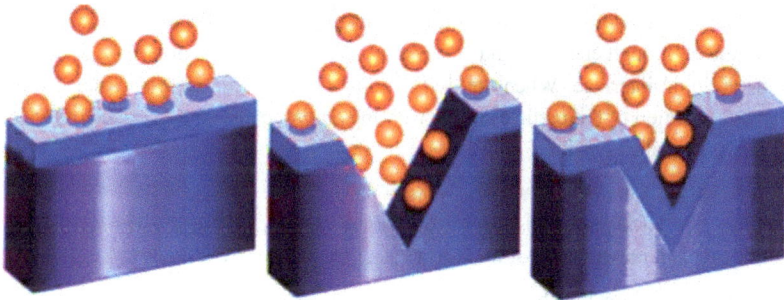

Figure 4.1. Formation of the thin protective film on the surface of stainless steel that gives it corrosion resistance and chemical inertia [15].

In addition to its resistance to corrosion and chemical inertia, stainless steel has many other properties that make it ideal for use in equipment in the food industry. It is easy to process it in such a way that it presents smooth surfaces with low roughness. Its hardness allows it to maintain this smoothness of the surface over time. The smoothness of the surface is extremely important because it is very well demonstrated that the adhesion of food to the surface (which is undesirable in the food industry) increases with increased roughness. Having the possibility to process the stainless-steel surface to extremely good smoothness, it will remain clean, it will not be loaded with adherent layers of food that will accumulate in certain areas of the processing equipment and that will bring an increased risk of hygiene [16].

Stainless steel is resistant to the extremely varied temperatures that are used in the food industry. It is not altered by heating, boiling, baking, cooling, freezing. Very important is the fact that the physical properties of stainless steel give it a good machinability. Depending on the composition, the stainless steel is processed well by deformation, cutting, welding well, which makes the equipment in the food industry can be made in conditions of economy and with excellent functionality. There is a guarantee, given by the material, that they will withstand impact, fatigue, wear, abrasion and corrosion [17].

It is interesting to note that when the equipment made of stainless steel has completed its life cycle and needs to be replaced, stainless steel still does not exhaust its service life, it can be reused. Today, a new equipment produced contains, on average, 60% recycled stainless steel.

A final property that we emphasize here and that ensures the success of stainless steel in the exposure part of the food and in the domestic environment is its extremely pleasant appearance and the feeling of cleanliness it offers [18].

Corrosion is a complex process of destruction of a material caused by the action of the environment through chemical and electrochemical processes that take place at the metal-gas or metal-liquid interface. When the destruction of the metallic material occurs through the mechanical action of the particles in the working environment on the surface of the material, the process is called erosion.

The corrosion of metallic materials, besides the fact that it reduces the operating safety of machines and installations, also causes important metal losses. It is estimated that annually quantities that represent 10% of the world's steel production are lost through corrosion.

It is good to understand that a metal corrodes in a working environment as a result of electrochemical oxidation and reduction reactions, called conjugated electrochemical reactions. The speed of these reactions depends on the electrode potential. The oxidation reaction corresponds to the dissolution of the metal (M) that passes into the solution in the state of ions, after a reaction of the general form:

$$M \rightarrow M^{n+} n \cdot e - \tag{4.1}$$

The reduction reaction consists in reducing an oxidizing agent capable of receiving the results from the ionization of the metal and can be written in the general form:

$$O_x + p \cdot e - \rightarrow R_{ed} \tag{4.2}$$

in which: O_x is the oxidized form of a constituent in the working environment, Red is the reduced form of the same constituent.

Therefore, the corrosion phenomenon consists of an exchange of ions between the metal and the working environment and any modification of this exchange is very important. During corrosion, a certain potential is established on the metal, the value of which is between the standard values of the partial electrode potentials. The displacement of the two equilibrium potentials to the corrosion potential is a consequence of the phenomenon of polarization of electrode processes, a phenomenon that occurs instantly, with the appearance of metal contact – work environment. Polarization is the displacement of the equilibrium potential to a new value, under the action of passing its own corrosion electric current, produced by the electrochemical reactions conjugated by the electrode [19].

Polarization is a brake on the development of corrosion. Due to polarization, the corrosion rate can decrease tens or hundreds of times compared to the initial value. After a period of attack, thin layers of pure metal are formed, of chemical compounds (salts or oxides) that protect the metal against the attack of the working environment, that is, the passivation of the metal takes place [20].

In the case of stainless steels, iron, nickel, chromium is of interest in passivation of the component elements.

• iron becomes passive in concentrated nitric acid, in acidic or neutral solutions of silver or copper nitrate, in chromic acid and its salts, in aerated alkaline solutions; the determination of the passivity of the iron depends on the conditions of the formation of the protective layer; this layer is easily destroyed by the ions of Cl⁻;

• nickel becomes passive in a manner similar to iron, but its passivity is less stable and more difficult to reproduce;

• chromium is more attackable than iron; the sensitivity of chromium to passivation is very high, but stable and corresponds to a higher positive potential than silver and copper; passivity of chromium is a stable state in the air, which makes it interesting for industrial uses; in iron-chromium alloys, chromium contributes to the passivation of iron, due to the high tendency to absorb electrons; iron can become apparently passive by losing at least one electron per atom; chromium, having incompletely the 3d electronic layer, can absorb five electrons, that is, it can passivize five electrons coming from the 3d layer of iron; this proportion corresponds to 15.7% Cr in weight, which explains why the content of over 12% Cr gives steel a good resistance to corrosion [21].

As a result of these considerations, perhaps difficult to understand by laypeople in the science of materials, we emphasize once again that the corrosion resistance of stainless steels is given by the formation on their surface, spontaneously, of a passive film of chromium oxides, resulting from the interaction between oxygen in the working environment and chromium in the steel, which is absolutely necessary to be more than 10.5%. [22].

Stainless steels, however, have various forms of corrosion, which depend on: the type of steel, the working environment and the operating conditions. Various types of corrosion are dependent on the mechanism of the material destruction process [23].

The types of corrosion encountered in stainless steels are:

1. General and local corrosion

2. Intercrystalline (intergranular) corrosion

3. Point corrosion (pitting)

4. Live corrosion

5. Cavernous (cracking) corrosion

6. Galvanic corrosion.

General and local corrosion is characterized by a relatively uniform degradation of the entire metal surface exposed to corrosive environment. Nitric acid and sulfuric acid are examples of agents that cause corrosions of this type. The general corrosion of stainless steels occurs almost exclusively in highly acidic or alkaline environments [24, 25].

Intercrystalline (intergranular) corrosion is an attack that progresses to the limits of the austenite grains. It is caused by the fact that, under certain temperature conditions, chromium carbide precipitates at the limit of the crystalline grains, forming around them a small area, depleted in chromium. The material, in these areas, is no longer stainless and it is quickly attacked in contact with an acidic electrolyte. There are solutions to diminish the sensitivity to intercrystalline corrosion [26]:

- the use of a steel with a low carbon content;

- the use of a steel stabilized with titanium or niobium;

- repetition of the heat treatment of the solution.

Point corrosion (pitting) is caused by a series of corrosive agents (halogenated salts Cl- , Br- and I-) which have the property of favoring local breaks of the superficial film for the protection of passivated steels. At each breaking point, a micro-anode is formed in which the electrical density is very high compared to the cathodic surface, relatively very high. This potential difference produces a punctate attack, which develops rapidly in depth [27].

Live corrosion is specific to austenitic stainless steels and occurs when the working environment has a specific action, when there is a certain sensitivity of the steel and when it is subjected to mechanical traction stresses. Traction voltages, caused by forces outside steel or by residual tensions following mechanical-thermal processing, are the most dangerous for the occurrence of live corrosion. Traction tensions cause slips in austenitic grains and sliding thresholds are active elements of corrosion [28].

Cavernous (cracking) corrosion occurs in stainless steels especially in the presence of chlorides. Cavernous corrosion occurs under the rubber sealing layers or on the assembly surfaces, in which case the attack can be avoided by applying a sealing mastic to the edges of the gasket. Cavernous corrosion was manifested not only on the contact surface between non-metallic gaskets and stainless steel, but also on the contact surfaces of two stainless steels [29].

Galvanic corrosion occurs if two different metals or a metal and an electroconducting material form a closed electrical circuit in the presence of an electrolyte. Galvanic corrosion can be prevented by avoiding direct contact between two metals that have large differences in electrical potential [30-32].

4.2. Classification, influence of alloying elements on stainless steel alloys

Stainless steels are iron-carbon alloys, which contain as main alloying element minimum 10.5% chromium. It is well known that ordinary steels exposed to unsuitable environmental conditions rust, on their surface appears a layer of rust made of oxides that do not protect the material, but develop in depth, in time leading to the compromise of the respective piece. But if there is enough chrome in the steel composition, more than 10.5%, the layer of oxides on the surface has special properties, is extremely thin, protects the piece in extremely varied and hostile environments and makes itself very easy when removed. This layer is the one that gives the steels the character of "stainless" and its self-creation gives superiority over the usual steels that for the purpose of their protection are covered with galvanically deposited layers of zinc or cadmium or are painted, and these coatings once damaged are restored only by the intervention of the humans, involving new costs [33, 34].

Along with iron, carbon and chromium, other elements are added to the composition of stainless steels that improve their properties or give them new ones. The increased carbon content gives hardness and mechanical strength. The addition of nickel stabilizes the austenitic structure, the material is nonmagnetic and less fragile at low temperatures.

Manganese acts the same as nickel, giving the material the same properties, but at a lower cost. Molybdenum increases corrosion resistance and high temperature.

From the point of view of their crystalline structure, determined by the chemical composition, stainless steels are classified into five types (Figure 4.2):

• **austenitic and superautistic:** contain a maximum of 0.15% carbon, minimum 16% chromium and enough nickel and/or manganese to stabilize the austenitic structure; the addition of nickel to stainless steels improves their deformability and weldability; an addition of 8...12% nickel allows the stainless steel to be rolled, pressed, pressed, drawn and also increases the corrosion resistance;

• **ferritics:** contain 10,5....18% chromium and approx. 0,05% carbon;

• **martensitics:** contain approx. 13% chromium and high percentages of carbon (even over 1%); are the cheapest stainless steels, but are difficult to deform and weldable; • duplex: have extremely high content of chromium (over 22%) and approx. 3% molybdenum; they resist in the most corrosive environments;

• **precipitation-hardenable:** they contain as an alloying element copper (which improves the resistance to acids) and niobium (which reduces corrosion in the weld area); they are expensive stainless steels, they have high processing costs, but they combine the remarkable corrosion resistance of austenitic steels with the excellent mechanical properties of martensitic steels [35, 36].

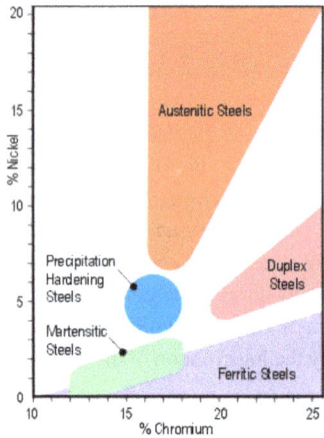

Figure 4.2. Types of stainless steels according to the content of chromium and nickel [37].

Table 4.2. Comparison of properties of different types of stainless steels [1-4].

Type of stainless steels	Corrosion resistance	Ductility	High temperature resistance	Low temperature resistance
Austenitic	high	very high	very high	very high
Ferritic	average	average	high	Low
Martensitic	average	Low	Low	Low
Duplex	very high	average	Low	average
Hard-liable by precipitation	average	average	Low	Low

Table 4.3. Comparison of properties of different types of stainless steels [1-4].

Type of stainless steels	Magnetic response	Possibility of hardening	Speed of hardening	Weldability
Austenitic	no, in general	by cold processing	very high	very good
Ferritic	yes	Not	average	Low
Martensitic	yes	by hardening+ rebound	average	Low
Duplex	yes	not	average	hi
Hard-liable by precipitation	yes	by aging	average	hi

When choosing the type of stainless steel to be used in a particular application, the following must be taken into account: corrosion resistance, physical and mechanical properties, machining possibilities and costs. The main selection criterion must be the corrosion resistance in the environment in which the stainless-steel element will work, and then the fulfillment of the other criteria will be followed, at a lower cost. It should be noted that the criterion of the minimum cost of material is not essential here, but that of the minimum total cost involved in operating throughout the lifetime. Stainless steel elements, even if they are more expensive in terms of material and processing costs, will avoid a whole series of costs related to repair, re-coating, repainting, failure in operation and the need for replacement, involved by other materials that do not resist corrosion [38].

In table 4.2 and table 4.3, general comparative indications are given concerning the properties of the main types of stainless steel. Table 4.4 summarizes the advantages and disadvantages of each type of stainless steel and gives, for a better understanding, examples of steel brands.

Table 4.4. Advantages and disadvantages of different types of stainless steels [1-4].

Type of stainless steels	Advantages	Disadvantages	Example EN(AISI)
Austenitic	no, in general	by cold processing	1.4301(304)
			1.4401 (316)
Ferritic	The most used, good corrosion resistance, resistance in cryogenic conditions, excellent deformability, good weldability	Hardening can limit deformability, low resistance to fatigue corrosion	1,400 (410S)
			1.4016 (430)
			1.4749 (446)
Martensitic	Low cost, good deformability	Lower corrosion resistance and deformability than austenitic stainless steels	1.4021 (420)
			1.4057 (431)
Duplex	High hardness and mechanical strength, hardenable by heat treatment, low cost	Limited corrosion resistance compared to austenitics, limited deformability compared to ferritics, low weldability	1.4501
			1.4462
Hard-liable by precipitation	Hardenable by heat treatment, better corrosion resistance than martensitics	Hardly available, expensive, corrosion resistance, deformation and restricted weldability compared to austenitics	1.4542 (630)
			1.4568 (631)

Taking into account the structure and properties of different types of stainless steels, some general indications can be drawn for choosing the appropriate type of material according to the requirements of each application, indications summarized in table 4.5.

Table 4.5. Choosing the right type of stainless steel according to the requirements of each application [1-4].

Requirement	The type of stainless steel that can be selected
Corrosion resistance	The selection depends on the environment in which the item will work.
High temperature resistance	Austenitic steels, especially those with a high chromium content, often with a high content of silicon, nitrogen, e.g., brand 1.4845 (310). Ferritic stainless steels with a high chromium content may also be used, e.g., 1.4749 (446).
Low temperature resistance	Austenitic steels, which have excellent toughness at extremely low temperatures.
Operation in magnetic fields	Austenitic steels, because they have low magnetic permeability, those with high nickel content, e.g. 1.4401 (316) or 1.4845 (310) are guaranteed to be nonmagnetic even after cold deformation
High mechanical strength	Martensitic steels and precipitation-hardenable steels

Stainless steel, like all steels by the way, are alloys with an iron base. Chromium is added to improve corrosion resistance by forming a layer of chromium oxide on the surface; it requires at least 17% chromium in the chemical composition of stainless steel. Carbon and nickel are used in the alloy as elements that increase strength. The most commonly used type of stainless steel used for implants is the type 316L (ASTM F 138), which contains 17-19% Cr, 13-15.5% Ni, and less than 0.03%C [39].

The lying with chromium generates a self-regenerating protective oxide that resists perforation and has a high degree of electrical resistivity, thus ensuring a very good protection against corrosion; the formation of the passive layer of chromium oxide is facilitated by the immersion of the alloy in a concentrated solution of nitric acid. Nickel increases corrosion resistance and eases the manufacturing process. The addition of molybdenum improves the resistance to corrosion of the "pitting" type (Table 4.6) [40].

ASTM F 4 surface passivation specifications are applicable to stainless steel-based alloys. Of the alloys used to make implants, this alloy is the most susceptible to corrosion in crevasse and "pitting" and it should be taken into account the obtaining and use of surface passivation. Due to the fact that this alloy contains nickel as the main alloying element, it should be avoided the use in patients allergic or hypersensitive to nickel. Also, if a stainless-steel implant is used in a surgical operation, the preceding procedures provide for repasivation in order to obtain an oxidized surface to reduce biodegradation in vivo.

Fe-based alloys have galvanic potentials and corrosion characteristics that can lead to the formation of galvanic couples and to biocorrosion, if they are interconnected with implants of Ti, Co, Zr or C. For example, if a support made of a noble alloy or a base metal of the alloy reaches the connecting ends of a stainless-steel implant, simultaneously with those of a titanium implant, an electric current will form through the tissue.

Table 4.6. Stainless steels [1].

Material	F 138	F 745	F 1314	F 1586
Density (gr/cm^3)	7,9	-	7,98	-
E (GPa)	200	-	200	200
Hardness (HV)	350	-	205	365
$\sigma_{0.2\%}$ (MPa)	690	207	380	975
UTS σ (MPa)	860	483	690	1090
Elongation (%)	12	30	35	14,5

If used independently, and the alloys were not in electrical contact, galvanic torque would not exist and each device could operate independently. The implantation of long-term devices has shown that, when used correctly, the alloy works well in vivo. Clearly, the mechanical properties and cost characteristics of this alloy offer advantages in clinical applications.

The first stainless steel used for metal implants was 18-8 steel (type 302), which was tougher than vanadium steel and more resistant to corrosion [41].

Vanadium steel is no longer equipped at present due to its indicated resistance to corrosion. Subsequently, the 18-8 stainless steel with molybdenum was introduced, in order to improve the corrosion resistance in salt water. This alloy is known south as the name of 316 stainless steel. In 1950, carbon content was reduced from 0.08% gr. At 0.03% gr. To obtain a better resistance in chlorine solution, the steel becoming 316 L [42].

Chromium is the majority component of corrosion-resistant stainless steels. The minimum effective concentration of Cr is 11% gr. Chromium is a reactive element, but its alloys can pass by having excellent corrosion resistance. In table 4.7 are presented the chemical compositions of steels 316 and 316 L:

Table 4.7. Chemical composition of stainless steels 316 and 316 L [1-4].

Bloke	%C	%Cr	%Ni%	%Mo%,	%Mn	%Si	%P	%S%;	%Fe %
316	0,006	17..20	12..14	2..4	2	0,75	0,03	0,03	rest
316 L	0,002	17..200	12.14	2..4	2	0,75	0,03	0,03	rest

These stainless steels are most used for metal implants. They are not hardenable by heat treatment, but they are hardenable by cold processing. The presented group is nonmagnetic having the highest corrosion resistances compared to other stainless steels, and the alloying with molybdenum increases the resistance to pitting corrosion in salt water.

For the manufacture of implants, the American Society of Materials – ASTM recommends 316 L steel rather than 316.

Although it is known as a toxic element, which in many situations generates the body's responses in its presence, this are in the chemical composition in the form of an alloying element that stabilizes austenitic at room temperature and, in addition, increases corrosion resistance. The stability of austenitic at such low temperatures is also influenced by the content of Cr [43].

Austenitic stainless steels undergo a hardening process in operation, which is why they cannot be cold processed without the application of intermediate heat treatments. However, heat treatments must not cause the appearance of chromium carbides (CCr4) at the limit of grain, carbides that can cause the generation of the corrosion phenomenon.

That is why austenitic stainless-steel implants do not weld in the known regimen. As a result of the thermal treatments, a distortion of the composition may occur, which can be highlighted by rigorously controlling the uniformity of the heating. Also, after the heat treatment appears an unwanted layer of oxide that is removed by chemical or mechanical methods (sandblasting).

In dentistry, stainless steels are used as threads or frames for partial teeth (in the form of plastic deformed products) and crowns or support for molten porcelain (in the form of molded products).

4.3. Applications in medicine

The principal hardened steel utilized as a material for making an embed was 18-8 (type 302 in the advanced characterization), which has a higher obstruction than vanadium and is significantly more impervious to consumption. Vanadium-based steel is at this point not utilized in inserts since its erosion obstruction is lacking. In this manner, the 18%Mo treated steel started to be utilized. It contains molybdenum to further develop erosion opposition in salt water. This combination started to be known as tempered steel type 316. During the 1950s, the measure of carbon in the treated steel type 316 was decreased from 0.08% load to a limit of 0.03% load for more prominent consumption obstruction in chloride; this new amalgam was known as the kind 316L [44-46].

Table 4.8. Categories of stainless steels used in medical applications [47].

Type of material	Applications	Examples
Martensitic stainless steel	Dental and surgical instruments	Dental cutters, dental chisel, curettles, dental wires
Stainless ferritic steel	Surgical instruments	Bolt of guidance and buckle.
Austenitic stainless steel	Non-implantable medical equipment Short-term implants Hip Replacement Implants	Guide bolt, hypodermic needles, steam sterilizer

Stainless steels are used as biomaterials (Table 4.8) only to make temporary medical implants, such as fixing screws and orthopedic rods for fixing fractures, because the passive (corrosion-resistant) layer of these steels is not as robust as in the case of titanium alloys.

The overwhelming majority of surgical instruments is made of metal materials from the category of austenitic and martensitic stainless steels, but also of titanium alloys, etc., materials that possess good corrosion resistance, adequate mechanical properties and a corresponding biocompatibility.

Stainless steel surgical alloys (Table 4.9) have a long history of use for the manufacture of orthopedic devices and dental implants. These alloys, like titanium alloys, are most commonly used in forged and heat-treated state, which gives them high hardness and ductility.

The service life of surgical instruments is generally low due to the change in the surface properties of metal instruments following faulty maintenance by the medical staff performing the decontamination and sterilization operations (they do not have knowledge of the science of materials), the improper use of chemical reagents or a low quality of water used in decontamination processes [48].

Table 4.9. The composition of stainless-steel type 316L used for surgical implants [47].

Element	Composition (%weight)
Carbon	0.030 max.
Magnesium	2,00 max.
Phosphorus	0,025 max.
Sulfur	0.010 max.
Silicone	0.75 max.
Chromium	17.00-19.00
Nickel	13.00-15.00
Molybdenum	2.25-3.00
Nitrogen	0.10 max.
Copper	0.50 max.
Iron	balance

As a result of these surface changes in the metal instrument, serious cases of infections and allergic reactions of patients may occur, under the conditions that most of these tools come into contact with human tissues.

The solutions identified to improve these surface properties of surgical instruments (Figure 4.3) made of metal materials consist of deposits of protective layers [49].

Figure 4.3. Surgical instruments [50].

Surgical steel refers to a variety of stainless steels used in biomedical applications. Sometimes called surgical stainless steel, surgical steel does not have a formal definition. However, the types of stainless steel with the highest level of corrosion resistance are also those intended for biomedical use. Some commonly accepted types of surgical steel include 316 austenitic steel and 440 and 420 martensitic steels. It may be useful to consider surgical steel as the stainless steel with the highest corrosion resistance.

Surgical steels are those with the highest corrosion resistance and are intended for biomedical applications. Compared to other types of steel, stainless steel is usually the most expensive.

References

[1] Davis, J.R. Introduction to stainless steels, Alloy Digest Sourcebook: Stainless steels. *ASM International* **2000**, 584.

[2] Xiao, M.; Chen, Y.M.; Biao, M.N.; Zhang, X.D.; Yang, B.C. Bio-functionalization of biomedical metals. *Mater. Sci. Eng. C Mater. Biol. Appl.* **2017**, *70,* 1057–1070. https://doi.org/10.1016/j.msec.2016.06.067

[3] Medilanski, E.; Kaufmann, K.; Wick, L.Y.; Wanner, O.; Harms, H. Influence of the surface topography of stainless steel on bacterial adhesion. *Biofouling* **2002**, *18(3),* 193–203. https://doi.org/10.1080/08927010290011370

[4] Zardiackas, L. Stainless steel for implants. *Wiley Encyclopedia of Biomedical Engineering* **2006**. https://doi.org/10.1002/9780471740360.ebs1136

[5] Chen, Q.Z.; Thouas, G.A. Metallic implant biomaterials. *Mater. Sci. Eng. R. Rep.* **2015**, *87,* 1–57. https://doi.org/10.1016/j.mser.2014.10.001

[6] Hermawan, H.; Ramdan, D.; Djuansjah, J. Metals for biomedical applications, in: R. Fazel-Rezai (Ed.). *Biomedical Engineering – From Theory to Applications*, **2011**, 411–430. https://doi.org/10.5772/19033

[7] Oberringer, M.; Akman, E.; Lee, J.; Metzger, W.; Akkan, C.K.; Kacar, E.; Demir, A.; Abdul-Khaliq, H.; Puetz, N.; Wennemuth, G.; Pohlemann, T.; Veith, M.; Aktas, C. Reduced myofibroblast differentiation on femtosecond laser treated 316LS stainless steel. *Mater. Sci. Eng. C Mater. Biol. Appl.* **2013**, *33(2),* 901–908. https://doi.org/10.1016/j.msec.2012.11.018

[8] Park, J.; Kim, D.J.; Kim, Y.K.; Lee, K.H.; Lee, H.; Ahn, S. Improvement of the biocompatibility and mechanical properties of surgical tools with TiN coating by PACVD. *Thin Solid Films* **2003**, *435(1–2)*, 102–107. https://doi.org/10.1016/S0040-6090(03)00412-7

[9] Hosseinalipour, S.M.; Ershad-langroudi, A.; Hayati, A.N.; Nabizade-Haghighi, A.M. Characterization of sol-gel coated 316L stainless steel for biomedical applications. *Prog. Org. Coat.* **2010**, *67(4)*, 371–374. https://doi.org/10.1016/j.porgcoat.2010.01.002

[10] Kheirkhah, M.; Fathi, M.; Salimijazi, H.R.; Razavi, M. Surface modification of stainless steel implants using nanostructured forsterite (Mg2SiO4) coating for biomaterial applications. *Surf. Coat. Technol.* **2015**, *276*, 580–586. https://doi.org/10.1016/j.surfcoat.2015.06.012

[11] Okner, R.; Domb, A.J.; Mandler, D. Electrochemically deposited poly(ethylene glycol)-based sol-gel thin films on stainless steel stents. *New J. Chem.* **2009**, 33(7), 1596–1604. https://doi.org/10.1039/b901864f

[12] Mahapatro, A.; Johnson, D.M.; Patel, D.N.; Feldman, M.D.; Ayon, A.A.; Agrawal, C.M. Surface modification of functional self-assembled monolayers on 316L stainless steel via lipase catalysis. *Langmuir* **2006**, *22(3)*, 901–905. https://doi.org/10.1021/la052817h

[13] Kaufmann, C.R.; Mani, G.; Marton, D.; Johnson, D.M.; Agrawal, C.M. Long-term stability of self-assembled monolayers on 316L stainless steel. *Biomed. Mater.* **2010**, *5(2)*. https://doi.org/10.1088/1748-6041/5/2/025008

[14] Choudhary, G.; Singh, G. A review of corrosion behaviour analysis studies of different stainless steel grades in distinct environments. *Int. J. Latest Trends Eng. Technol. Spec. Issue* **2017**, 178–183.

[15] Information on https://www.novameta.lt/en/products/cleaning-care-and-maintenance-of-stainless-steel. Available online: (accessed on September 10, 2021).

[16] Mozafari, M.; Salahinejad, E.; Sharifi-Asl, S.; Macdonald, D.D.; Vashaee, D.; Tayebi, L. Innovative surface modification of orthopaedic implants with positive effects on wettability and in vitro anti-corrosion performance. *Surf. Eng.* **2014**, *30(9)*, 688–692. https://doi.org/10.1179/1743294414Y.0000000309

[17] Gopi, D.; Ramya, S.; Rajeswari, D.; Kavitha, L. Corrosion protection performance of porous strontium hydroxyapatite coating on polypyrrole coated 316L stainless steel. *Colloids Surf. B. Biointerfaces* **2013**, *107*, 130–136. https://doi.org/10.1016/j.colsurfb.2013.01.065

[18] Macionczyk, F.; Gerold, B.; Thull, R. Repassivating tantalum/tantalum oxide surface modification on stainless steel implants. *Surf. Coat. Technol.* **2001**, *142*, 1084–1087. https://doi.org/10.1016/S0257-8972(01)01096-9

[19]   Soya, M.; Yoshioka, T.; Shinozaki, K.; Tanaka, J. Effect of oxide layers of Ni-free stainless-steel on silane coupling agent immobilization. *Mater. Trans.* **2009**, *50(6)*, 1318–1321. https://doi.org/10.2320/matertrans.ME200827

[20]   Bastarrachea, L.J.; Goddard, J.M. Development of antimicrobial stainless steel via surface modification with N-halamines: characterization of surface chemistry and N-halamine chlorination. *J. Appl. Polym. Sci.* **2013**, *127(1)*, 821–831. https://doi.org/10.1002/app.37806

[21]   Zhang, F.; Kang, E.T.; Neoh, K.G.; Wang, P.; Tan, K.L. Surface modification of stainless steel by grafting of poly(ethylene glycol) for reduction in protein adsorption. *Biomaterials* **2001**, *22(12)*, 1541–1548. https://doi.org/10.1016/S0142-9612(00)00310-0

[22]   Harvey, J.; Bergdahl, A.; Dadafarin, H.; Ling, L.; Davis, E.C.; Omanovic, S. An electrochemical method for functionalization of a 316L stainless steel surface being used as a stent in coronary surgery: irreversible immobilization of fibronectin for the enhancement of endothelial cell attachment. *Biotechnol. Lett.* **2012**, *34(6)*, 1159–1165. https://doi.org/10.1007/s10529-012-0885-8

[23]   Skovager, A.; Whitehead, K.; Wickens, D.; Verran, J.; Ingmer, H.; Arneborg, N. A comparative study of fine polished stainless steel, TiN and TiN/Ag surfaces: adhesion and attachment strength of Listeria monocytogenes as well as anti-listerial effect. *Colloids Surf. B. Biointerfaces* **2013**, *109*,190–196. https://doi.org/10.1016/j.colsurfb.2013.03.044

[24]   Xiao, Y.L.; Zhao, L.; Shi, Y.F.; Liu, N.; Liu, Y.L.; Liu, B.; Xu, Q.H.; He, C.L.; Chen, X.S. Surface modification of 316L stainless steel by grafting methoxy poly(ethylene glycol) to improve the biocompatibility. *Chem. Res. Chin. Univ.* **2015**, *31(4)*, 651–657. https://doi.org/10.1007/s40242-015-5027-0

[25]   Manivasagam, G.; Dhinasekaram, D.; Rajamanicham, A. Biomedical implants: corrosion and its prevention - a review. *Recent Patents on Corrosion Science* **2010**, 40–54.

[26]   Garcia, A. Surface modification of biomaterials, in: Atala, A.; Lanza, R.; Thomson, J.; Nerem R. (Eds.). *Principles of Regenerative Medicine* **2011**. https://doi.org/10.1016/B978-0-12-381422-7.10036-7

[27]   Latifi, A.; Imani, M.; Khorasani, M.T.; Joupari, M.D. Electrochemical and chemical methods for improving surface characteristics of 316L stainless steel for biomedical applications. *Surf. Coat. Technol.* **2013**, *221*, 1–12. https://doi.org/10.1016/j.surfcoat.2013.01.020

[28]   Bhuyan, A.; Gregory, B.; Lei, H.; Yee, S.Y.; Gianchandani, Y.B. Pulse and DC electropolishing of stainless steel for stents and other devices. *IEEE Sensors*, **2005**, *1-2*, 314–317.

[29]   Talha, M.; Behera, C.K.; Sinha, O.P. A review on nickel-free nitrogen containing austenitic stainless steels for biomedical applications. *Mater. Sci. Eng. C Mater. Biol. Appl.* **2013**, *33(7)*, 3563–3575. https://doi.org/10.1016/j.msec.2013.06.002

[30]   Shih, C.C.; Shih, C.M.; Su, Y.Y.; Su, L.H.J.; Chang, M.S.; Lin, S.J. Effect of surface oxide properties on corrosion resistance of 316L stainless steel for biomedical applications. *Corros. Sci.* **2004**, *46(2)*, 427–441. https://doi.org/10.1016/S0010-938X(03)00148-3

[31]   Muller, R.; Abke, J.; Schnell, E.; Macionczyk, F.; Gbureck, U.; Mehrl, R.; Ruszczak, Z.L.; Kujat, R.; Englert, C.; Nerlich, M.; Angele, P. Surface engineering of stainless steel materials by covalent collagen immobilization to improve implant biocompatibility. *Biomaterials* **2005**, *26(34)*, 6962–6972. https://doi.org/10.1016/j.biomaterials.2005.05.013

[32]   Xie, D.; Wan, G.J.; Maitz, M.F.; Sun, H.; Huang, N. Deformation and corrosion behaviors of Ti-O film deposited 316L stainless steel by plasma immersion ion implantation and deposition. *Surf. Coat. Technol.* **2013**, *214*, 117–123. https://doi.org/10.1016/j.surfcoat.2012.11.012

[33]   Tavares, J.; Shahryari, A.; Harvey, J.; Coulombe, S.; Omanovic, S. Corrosion behavior and fibrinogen adsorptive interaction of SS316L surfaces covered with ethylene glycol plasma polymer-coated Ti nanoparticles. *Surf. Coat. Technol.* **2009**, *203(16)*, 2278–2287. https://doi.org/10.1016/j.surfcoat.2009.02.025

[34]   Liu, Y.; Cao, H.; Yu, Y.; Chen, S. Corrosion protection of silane coatings modified by carbon nanotubes on stainless steel. *Int. J. Electrochem. Sci.* **2015**, *10(4)*, 3497–3509.

[35]   Benvenuto, P.; Neves, M.A.D.; Blaszykowski, C.; Romaschin, A.; Chung, T.; Kim, S.R.; Thompson, M. Adlayer-mediated antibody immobilization to stainless steel for potential application to endothelial progenitor cell capture. *Langmuir* **2015**, *31(19)*, 5423–5431. https://doi.org/10.1021/acs.langmuir.5b00812

[36]   Latifi, A.; Imani, M.; Khorasani, M.T.; Joupari, M.D. Plasma surface oxidation of 316L stainless steel for improving adhesion strength of silicone rubber coating to metal substrate. *Appl. Surf. Sci.* **2014**, *320*, 471–481. https://doi.org/10.1016/j.apsusc.2014.09.084

[37]   Information on https://www.azom.com/article.aspx?ArticleID=470. Available online: (accessed on September 10, 2021).

[38]   Virtanen, S.; Milosev, I.; Gomez-Barrena, E.; Trebse, R.; Salo, J.; Konttinen, Y.T. Special modes of corrosion under physiological and simulated physiological conditions. *Acta Biomater.* **2008**, *4(3)*, 468–476. https://doi.org/10.1016/j.actbio.2007.12.003

[39]   Bohinc, K.; Drazic, G.; Abram, A.; Jevsnik, M.; Jersek, B.; Nipic, D.; Kurincic, M.; Raspor, P. Metal surface characteristics dictate bacterial adhesion capacity. *Int. J. Adhes. Adhes.* **2016**, *68*, 39–46. https://doi.org/10.1016/j.ijadhadh.2016.01.008

[40]   Rodriguez, J.; Munoz-Escalona, P.; Cassier, Z. Influence of cutting parameters and material properties on cutting temperature when turning stainless steel. *Rev. Fac. Ing. Univ. Cent. Venez.* **2011**, *26(1)*.

[41]   Bagherifard, S.; Slawik, S.; Fernandez-Pariente, I.; Pauly, C.; Mucklich, F.; Guagliano, M. Nanoscale surface modification of AISI 316L stainless steel by severe shot peening. *Mater. Des.* **2016**, *102*, 68–77. https://doi.org/10.1016/j.matdes.2016.03.162

[42]   McLucas, E.; Moran, M.T.; Rochev, Y.; Carroll, W.M.; Smith T.J. An investigation into the effect of surface roughness of stainless steel on human umbilical vein endothelial cell gene expression. *Endothelium* **2006**, *13(1)*, 35–41. https://doi.org/10.1080/10623320600660185

[43]   Mahajan, A.; Sidhu, S. Surface modification of metallic biomaterials for enhanced functionality: a review. *Mater. Technol.* **2017** 1–13.

[44]   Liu, X.; Yue, Z.; Romeo, T.; Weber, J.; Scheuermann, T.; Moulton, S.; Wallace, G. Biofunctionalized anti-corrosive silane coatings for magnesium alloys. *Acta Biomater.* **2013**, *9(10)*, 8671–8677. https://doi.org/10.1016/j.actbio.2012.12.025

[45]   Kang, C.K.; Lee, Y.S. The surface modification of stainless steel and the correlation between the surface properties and protein adsorption. *J. Mater. Sci. Mater. Med.* **2007**, *18(7)*, 1389–1398. https://doi.org/10.1007/s10856-006-0079-9

[46]   Yoshioka, T.; Tsuru, K.; Hayakawa, S.; Osaka, A. Preparation of alginic acid layers on stainless-steel substrates for biomedical applications. *Biomaterials* **2003**, *24 (17)*, 2889–2894. https://doi.org/10.1016/S0142-9612(03)00127-3

[47]   Williams D.F., Biofunctionality and Biocompatibility, Materials Science and Technology. *Medical and Dental Materials*, Weinheim, **1992**, *14*, 2-27.

[48]   Bombac, D.M.; Brojan, M.; Fajfar, P.; Kosel, F.; Turk, R. Review of materials in medical applications. *Materials and Geoenvironment* **2007**, *54(4)*, 471-499.

[49]   Geetha, M.; Singh, A.K.; Asokamani, R.; Gogia, A.K. Ti based biomaterials, the ultimate choice for orthopaedic implants - A review. *Mater. Sci.* **2009**, *54*, 397-425. https://doi.org/10.1016/j.pmatsci.2008.06.004

[50]   Information on https://www.meadmetals.com/blog/surgical-steel-vs-stainless-steel. Available online: (accessed on September 10, 2021).

Advanced Metallic Biomaterials
Materials Research Foundations **118** (2022)

Materials Research Forum LLC
https://doi.org/10.21741/9781644901779

CHAPTER 5

Biodegradables Alloys

Magnesium is the chemical element in the periodic table of elements that has the symbol Mg. Magnesium is the eighth element and the third metal after aluminum and iron in abundance in the earth's solid crust, forming about 2% of its mass. Magnesium is the third most abundant component in salts dissolved in seawater [1].

The name is of Greek origin, Magnesia being the name of a region in Thessaly. The Englishman Joseph Black first identified magnesium as an element in 1755, Sir Humphry Davy first obtained pure magnesium in 1808 from a mixture of magnesium oxide and HgO, and A.A.B. Bussy prepared it in a bound form in 1831. Magnesium is a basic earth metal and is hence just found in blend with different components. It is found in enormous stores of magnesite (magnesium carbonate), dolomite and different minerals like powder [2].

Extraction is performed by precipitating $MgCO_3$ from seawater with $CaCO_3$:

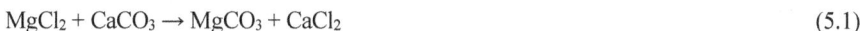

$$MgCl_2 + CaCO_3 \rightarrow MgCO_3 + CaCl_2 \tag{5.1}$$

The insoluble precipitate is filtered off, treated with HCl and an $MgCl_2$ concentrate is obtained from which extracts magnesium by electrolysis:

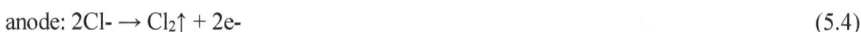

$$MgCO_3 + 2HCl \rightarrow MgCl_2 + CO_2 + H_2O \tag{5.2}$$

$$\text{cathode: } Mg_2+ + 2e\text{-} \rightarrow Mg \tag{5.3}$$

$$\text{anode: } 2Cl\text{-} \rightarrow Cl_2\uparrow + 2e\text{-} \tag{5.4}$$

Magnesium is present in large quantities in the human body representing the second cation encountered intracellularly, after potassium. It plays an important role in regulating neuromuscular function, being involved in many metabolic reactions and biological mechanisms. It is a key factor in the ribosomal mechanism of transcription of RNA-encoded genetic information into polypeptide structures. At the same time, it forms a complex with adenosine 5-triphosphate being absolutely necessary for maintaining its metabolism. It is a cofactor of intracellular transporters and enzymatic functions, regulating the function of over 300 enzymes. It is involved in maintaining the structure of the cell membrane intact and also in the function of the endocrine system through its role in regulating the level of parathyroid hormone [3].

The total amount of magnesium in the body is about 25 g (1000 mmol) with a normal plasma level of 1.7 - 2.4 mg / dl (1.5 - 2.0 mEg / L). The vast majority of magnesium is distributed in the bone (60%) having an important role in the development of bone tissue, contributing to its development and its mechanical strength. Muscle tissue is the next storage level of about 20% at this level. The remaining 20% are distributed in the other soft tissues. Magnesium is stored almost entirely at the cellular level, only 1% being extracellular, and of this 20% is bound to proteins and 80% is found in ionized form or forms complexes with other ions (phosphates, oxalates, nitrates) [4].

The daily food intake is normally about 360 mg (15 mmol). At the intestinal level the absorption varies between 25 and 75%, the amount absorbed at this level being about 120 mg. Of this, 20 mg is lost through gastrointestinal excretion, leading to a net absorption of approximately 100 mg. Magnesium excretion is performed by the renal system, but 60-70% of the filtered Mg is resorbed at the level of Henle's loops, and 5-10% is absorbed in the collecting tubes. Thus, renal excretion of Mg remains about 25%. The regulation of the Mg level in the body is not regulated by the endocrine system as in the case of other ions. The main factor that regulates the level of magnesium is its serum level. Thus, hypomagnesemia will increase the digestive absorption of magnesium and its tubular resorption, while hypermagnesemia will have the opposite effect, inhibiting these processes [5].

In this way the serum level of magnesium is simple to regulate, the intoxication with magnesium by its overdose being difficult to obtain. However, in some cases, pathological conditions may occur that may induce changes in magnesium homeostasis. Thus, hypomagnesaemia may be associated with some pathological conditions, the most common of which are: chronic diarrhea, impaired intestinal absorption capacity, nephropathy, excessive use of diuretics, Barrter syndrome, pancreatitis. Clinically it is manifested by neurological signs (lethargy, confusion, muscle fasciculations, tetany, ataxia, nystagmus) or cardiac (sinus or supraventricular tachycardia, ventricular arrhythmia). Hypermagnesemia is rare, being recorded in cases greater than 2.2 mEg/L. It occurs either as a result of chronic nephropathy that leads to decreased excretion, or due to a massive intake of Mg, but usually in the absence of an associated pathological condition, the body has the ability to eliminate large amounts of magnesium. Thus, in order to be clinically manifested, in the vast majority of cases it is necessary to reach serum concentration values above 4 mEg/L. From a clinical point of view, between 4 - mEg/L there may be decreases in tendon reflexes, lethargy, migraines. Between 6 and 9 mEg/L the tendon reflexes are abolished, drowsy, hypotension with EKG changes and at values higher than 10 mEg/L paralysis, respiratory failure or even cardiac arrest can be installed [6, 7].

Advanced Metallic Biomaterials   Materials Research Forum LLC
Materials Research Foundations **118** (2022)   https://doi.org/10.21741/9781644901779

5.1. Properties

The main properties of magnesium are shown in table 5.1. Magnesium is an extremely light metal, the relative density of magnesium being 1.74g/cm³. The high strength-to-weight ratio makes magnesium alloys extremely useful for industrial applications that require low-weight materials, such as the automotive or aerospace industry.

Table 5.1. Physico-mechanical properties of magnesium [8, 9].

Properties	Value
Atomic number, Z	12
Atomic mass, A	24.305 kg/kmol
Electronic configuration	1s22s2p63s2
Atomic radius	1.6 Å
Ionic radius	0.78 Å
Spread in the earth's crust	2% (the eighth element)
Color	Silvery-white
Density	1738 kg/m³
Liquid density	1584 kg/m³
Melting temperature	650°C
Boiling temperature	1090°C
Latent melting heat	8.48 kJ/mol
Vaporization heat	128 kJ/mol
Electrical resistivity	$4.39 \cdot 10^{-8}$ Ω·m, at 20°C
Thermal conductivity	156 w/mK, at 300K
Coefficient of thermal expansion	$24.8 \cdot 10^{-6}$ K^{-1}, at 25°C
Young module	45 GPa
Transverse modulus of elasticity	17 GPa
Mohs hardness	2.5
Poisson's ratio	0.290
Mechanical resistence	260 MPa
Stable isotopes	24Mg(78.99), 25Mg(10), 26Mg(11.01)
The crystallographic system	compact hexagonal, a = 3.21 Å and c = 5.21 Å
Magnetism	paramagnetic

Magnesium (Figure 5.1) in the form of castings or semi-finished parts is not flammable. It only ignites at a temperature close to the melting point. Magnesium does not react with oxygen in the air at low temperatures. Magnesium in the form of wire or powder burns when ignited in air, giving rise to abundant smoke of magnesium oxide and emitting a blinding white light. It is noted that the physicochemical properties of magnesium worsen with increasing temperature. Pure magnesium is resistant to hydrofluoric and chromic acids, as well as to alkaline hydroxide solutions [10].

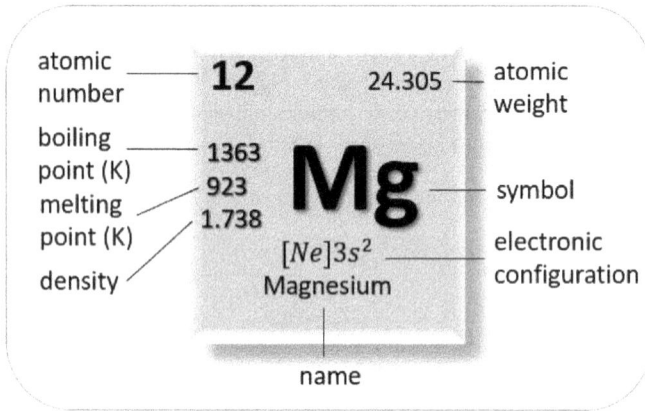

Figure 5.1. Chemical element Magnesium [11].

Absorbable materials should have physicochemical properties steady with the physiology of neighboring tissues. Their primary necessities are the capacity to be biocompatible and biodegradable.

By biocompatible is implied the capacity of an embed to be made out of material that doesn't initiate an organic reaction from contiguous tissues. It is portrayed by the shortfall of cancer-causing nature, immunogenicity, teratogenicity and harmfulness [12, 13].

Biodegradable is the capacity of a material to be synthetically disintegrated or to corrupt into compounds under the activity of natural variables (e.g., proteins). The biodegradable embed should be resorbed step by step, keeping up with its mechanical properties until it is presently not required, and afterward be resorbed without leaving any leftover mixtures in the body [14].

5.2. Classification, influence of alloying elements on magnesium properties

Magnesium and its alloys have been extensively studied and accepted as applications for the automotive and aerospace industries. These alloys contained various toxic elements, which were not suitable for biomedical applications. Thus, in recent years, new alloys applicable in medicine have been developed, Mg-based alloys with the following alloying elements: Ca, Si, Sr, Zn, Sn, Zr, Al and rare elements (Figure 5.2).

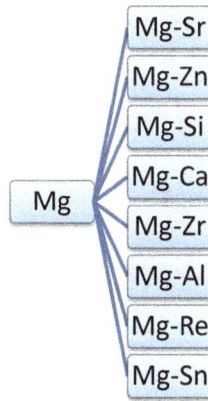

Figure 5.2. Classification of magnesium-based binary alloys [15, 16].

Magnesium alloys can be divided into three major groups [16]:

- pure magnesium (Mg) with traces of other elements,
- aluminum alloys,
- alloys that do not contain aluminum.

Typical alloys containing Al are: AZ91, AZ31 AE21, calcium-containing AZ alloys and LAE442, AZ31 and AZ91. These alloys also contain a low manganese (Mn) content. AE21 is composed of Mg, Al and rare earths (RE) and a small content of Mn. LAE442 alloy is based on AE42 alloy and contains Al, RE, Mn and in addition lithium (Li), being developed relatively recently as a magnesium alloy with a low density, good ductility and corrosion resistance superior to alloys in this category. Typical magnesium alloy systems without Al content are: WE, MZ, WZ and Mg-Ca binary alloys. The WE43 magnesium alloy, which contains Yttrium (Y), zirconium (Zr) and rare earths (RE), has been developed to improve creep resistance and stability at high temperatures. Magnesium-zinc (MZ) alloys have properties comparable to those of the ZM alloy system, which is a system used for industrial applications in the automotive industry [17].

The letter E used in the designation of a number of magnesium alloys represents rare earth elements (REE). In general, yttrium-containing alloys are marked with the corresponding letter, W. In industrial practice, magnesium alloys are hardened with a small amount of rare earths [18].

None of the above has been made to be used as a material for the execution of biodegradable implants. Due to the complex composition of the alloy, it is not possible to identify precisely whether the in vivo degradation analyzed in different ways can be attributed to the chemical elements in the composition, to an intermetallic compound or to the microstructure of the alloys.

Alloying is a general method used to improve the properties of pure magnesium. Alloying elements used mainly to obtain magnesium alloys are Al, Zn, Mn, Ca, Li, Zr and rare earths (RE) [35].

Depending on the selected alloying elements, but also depending on the production process, different structural characteristics of magnesium alloys are obtained (such as grain size and grain size distribution). Contrasted with the fundamental framework, the grain limit is a twisted region with numerous defects, high interior energy and any destructive assault in the unadulterated material typically assaults the grain limit first. The isolation of the alloying components from the grain limit happens relying upon the alloying components present and the picked hardening technique. Accordingly, the piece in the focal point of the grain will be unique in relation to that close to the grain limit which impacts the erosion conduct of magnesium amalgams [20, 21].

As an alloying element, aluminum (Al) can increase both the strength of the solid solution and the precipitates. Unfortunately, the $Mg_{17}Al_{12}$ phase in the Mg-Al system has a low melting point and cannot be used to improve resistance to high temperatures. In addition, increasing the aluminum content lowers the temperature of liquidus and solidus and increases the fluidity of alloys with a high aluminum content.

Zinc (Zn) is an alloying element often used for magnesium, being the second element after aluminum in terms of its properties. It helps remove the corrosive effect of impurities, such as iron or nickel, so it improves the corrosion resistance of magnesium alloys. Its concentration is usually limited to 3%. In addition, zinc is considered an essential micro-nutrient, as zinc deficiency can result in severe disruption of physiological functions in the human body [22].

Manganese (Mn) is mainly used to increase ductility. More significant is the development of Al-Mn intermetallic stages in magnesium amalgams that additionally contain aluminum, in light of the fact that these stages can extricate press and can be utilized to control the erosion of magnesium composites (because of the unfriendly impact of iron on consumption conduct). In smaller quantities, manganese contributes to the improvement of mechanical strength due to the increase in strength of the solid solution. It can also improve the fluidity of the alloy, but only if used in quantities of less than 2 wt.% [23].

Calcium (Ca) helps to increase the resistance of solid solution and precipitates. To a large extent, it can also act as a grain finishing agent and contribute to increasing the strength of the grain boundary. In Mg-Ca binary alloys, Laves-type intermetallic phases with the composition of Mg2Ca are formed, while in alloys with an aluminum content these Laves Al2Ca phases are formed at the beginning. Both phases improve the creep resistance due to the increase of the resistance of the solid solution, of the precipitations and due to the fixing of the grain limit. Using a higher amount of Ca (> 1 wt.%) can cause problems with hot casting [24].

Lithium (Li) is the only element that can change the structure of magnesium alloys from compact hexagonal to cubic with centered volume. Along these lines, it very well may be utilized to expand the pliability and versatility of magnesium compounds, however sadly negatively affects mechanical strength [25].

Zirconium (Zr) is commonly used to reduce grain size in magnesium alloys that do not contain aluminum, thus helping to increase the mechanical strength of these alloys. It likewise lessens the impact of iron pollutions on the consumption opposition of magnesium composites. In the Mg-Zr twofold combination framework, zirconium is successful in expanding the consumption opposition provided that it doesn't surpass 0.48% without the arrangement of zirconium-containing encourages.

Rare earths (RE) are usually introduced by using pre-alloys during the development of magnesium alloys, being alloys rich in hardening elements - Y, Ce, Nd (normally 50% cerium, 45% lanthanum, small amounts of neodymium and praseodymium) [42, 44]. In general, rare earths can be divided into two groups [26-28]:

• the main gathering contains components with high dissolvability in magnesium, like Yttrium (Y), gadolinium (Gd), terbium (Tb), dysprosium (Dy), holmium (Ho), erbium (Er), thulium (Tm), yterbium (Yb) and lutetium (Lu).

• the second group presents elements with limited solubility in magnesium: neodymium (Nd), lanthanum (La), cerium (Ce), praseodymium (Pr), samarium (Sm) and europium (Eu).

All rare earths can form complex intermetallic phases with aluminum and magnesium, intermetallic phases that act as obstacles to the movement of dislocations at high temperatures, thus contributing to the mechanical strength of the material.

In general, alloying elements influence the mechanical and physical properties of magnesium alloys in industrial applications. The influence of alloying elements and impurities on the properties and processing of magnesium alloys at ambient temperature is summarized in table 5.2.

Table 5.2. Pathophysiological and toxicological characteristics of alloying elements and impurities present in magnesium alloys [3, 18, 29].

Elements	Aspects of Pathophysiology / Toxicology
Magnesium	• normal level of magnesium in the blood: 0.73-1.06mmol/L • regulates metabolism, cell proliferation, protein synthesis • regulates the activity of over 350 proteins • stabilizes DNA and RNA • influences long-term cellular reactions
Calcium	• normal level of calcium in the blood: 0.919-0.993 mg/L • the most abundant mineral in the human body (1-1.1 kg) • stored in human hard tissues • role in the coagulation process • is regulated by the homeostasis of the skeletal, renal and intestinal mechanism • metabolic disorders → kidney stones
Aluminum	• normal level of aluminum in the blood: 2.1-4.8µg / L • stabilizes the alloying elements in titanium alloys • neurotoxic, affects the blood-brain barrier • risk factor in generating Alzheimer's disease • may cause muscle fibers to be destroyed • accumulation at the bone level → decreased osteoblast cell viability
Zinc	• normal level of zinc in the blood: 12.4-17.4 µmol / L • oligolement • is essential for the immune system • enzymatic agent in bones and cartilage • neurotoxic at high concentrations
Manganese	• normal level of manganese in the blood: <0.8 µg / L • essential trace element • important role in the metabolic circuit of lipids, amino acids, carbohydrates • influences the functions of the immune system, bone growth, coagulation BLOOD • neurotoxic at high concentrations (chronic manganese poisoning)
Lithium	• normal blood lithium level: 2-4ng / g • overdose causes kidney and lung dysfunction • possible teratogenic effects (produces malformations)
Rarely earths	• possible toxic effects on the liver and lungs • possible anticarcinogenic effect • incompletely elucidated toxic effects
Impurities	
Nickel	• normal blood level: 0.05-0.23 µg/L • strong allergen that can induce sensitivity to metals • carcinogenic and genotoxic
Beryllium	• toxic dose> 2 µg/m^3 • induces sensitivity to metals, strongly carcinogenic
Iron	• normal blood level: 5.0-17.6 g/L • regulated and stored metabolism
Copper	• normal blood level: 74-131 µmol/L • cellular toxic • high concentrations → degenerative neurological impairment

Thus, we can see that many of the alloys used successfully in metallurgy are not compatible with the human body, containing alloying elements with toxic potential. Magnesium-aluminum alloys, which are commonly used in metallurgy, have good mechanical strength and an optimal corrosion rate, are not an optimal choice for biomedical applications because the accumulation of aluminum in the body is associated with various disorders of the human nervous system due to neurotoxicity. this element. Although there are low limits for impurity concentrations compared to the range of physiological concentrations in the human body, elements such as beryllium and especially nickel should be removed from the alloys used for medical implants mainly due to carcinogenic effects [30-32].

On the other hand, calcium, zinc, manganese and rare earths are suitable for use in medical applications with minimal or no toxic effects.

Rare earth requires a special discussion. Due to their ability to harden magnesium alloys, increasing mechanical strength and having a good corrosion rate, they seemed an optimal solution for obtaining alloys with medical applications. But further studies have found that elements with high solubility in magnesium still have toxic effects on the human body.

However long the alloying components stay in the strong arrangement, they can be utilized to solidify the strong arrangement. Besides, the greater part of the given alloying components can respond with magnesium or can respond with one another to shape intermetallic stages, stages that add to the expansion of the strength of the compound by precipitation solidifying. Both solidifying of the strong arrangement and encourages work on the strength of the combination, yet lessen its malleability. As a rule, any alloying component adds somewhat to the decrease of grains, which fills in as a referred to solidifying instrument and as the fortifying of grain limits or the Hall-Petch impact [33, 34].

The impurities characteristic of magnesium alloys are: iron (Fe), copper (Cu), nickel (Ni) and beryllium (Be). In general, impurities can reach a total content of 0.3%, and very often these impurities are not presented in detail or even analyzed. The number of impurities depends on the composition of the alloy, the manufacturing technology and the evolution of the development of the alloy. It should be noted that most commercial magnesium alloys also contain impurities, which are not explicitly mentioned by the manufacturer. Thus, almost any commercial magnesium alloy containing aluminum is likely to contain manganese (0.4 - 0.6%) and silicon (up to 0.3%). Usually, beryllium is limited to 4 ppm, the amount of copper to 100 - 300 ppm, iron to 35 - 50 ppm and nickel should not exceed 20 - 50 ppm. The other chemical elements are considered as normal alloying elements and their limits are given together with their nominal content.

The combination is, be that as it may, restricted when planning magnesium composites for biomedical applications when the poisonousness of the alloying components is considered.

Most magnesium combinations that have been examined so particularly far as potential embed materials are very intricate and contain possibly poisonous alloying components. Magnesium compounds with an aluminum content, for instance, are not a favored decision for biomedical applications on the grounds that the amassing of aluminum is related with different problems of the human sensory system. Despite the fact that there are low cutoff points for debasement focuses contrasted with the scope of physiological fixations in the human body, components like beryllium and particularly nickel ought to be stayed away from [35, 36].

Curiously, magnesium particles are available in huge amounts in the human body and are associated with numerous metabolic responses and organic components. In addition, magnesium is fundamental for human digestion and is discovered normally in bone tissue. It is the fourth most plentiful cation in the human body, with an expected worth of 1 mole of magnesium put away in the body of an ordinary grown-up of 70 kg, with about portion of the physiological magnesium put away in bone tissue. This implies that magnesium can fill in as a biodegradable metallic material in the human body, in which magnesium can be progressively broken down, burned-through or consumed. Magnesium is a co-factor for some catalysts, and balances out DNA and RNA structures. The degree of magnesium in the extracellular liquid differs somewhere in the range of 0.7 and 1.05 mmol/L, in case homeostasis is kept up with by the kidneys and digestive organs [37, 38].

Since serum magnesium levels higher than 1.05 mmol/L can prompt muscle loss of motion, hypotension and respiratory pain, and heart failure happens in instances of high plasma focuses (6-7 mmol/L); the occurrence of hyper-magnesium is uncommon, because of the effective disposal of the component in the pee [39, 40].

Biodegradable metals will be metals that steadily erode in vivo, with the host organic entity having a proper reaction to erosion items, which will totally break down in the body after the mission of mending the encompassing tissue. In this manner, the significant part of biodegradable metals would be the fundamental metallic components, which can be processed by the human body, and show the pace of debasement in the human body [41-43].

Materials science classifies biodegradable metal materials into the following categories: pure metals, biodegradable metal alloys and biodegradable metal matrix composites.

The category of pure materials mainly includes a metallic element, with a level of impurity below the limits of commercial tolerance. The corrosion rate of these metals depends very much on the impurities found in the composition of the material.

The class of biodegradable metal amalgams involves biodegradable metals with substance sytheses (having at least one alloying components) and diverse trademark microstructures, contingent upon the technique for additional elaboration and handling. Given the worries for the biosafety of consumption items, the measure of alloying components should be controlled, as it isn't allowed for some compound components in the organization of these amalgams to cause pathophysiological and toxicological unfriendly impacts [44].

Table 5.3. The mechanical properties of some biodegradable metallic biomaterials concentrated up until this point [3, 50-52].

Metallic biomaterial (name, metallurgical state, chemical composition [wt%])	Density [g/cm³]	Flow limit [MPa]	Breaking stress [MPa]	Young's module [GPa]	Elongation [%]
Austenitic stainless steel type 316L, deformed and heat treated *(ASTM, 2003)** Fe, 16-18.5%Cr, 10-14%Ni, 2-3% Mo, 2%Mn, 1%Si, 0.03%C	8.00	190	490	190	40
Iron, deformed condition and heat treated *(Goodfellow, 2007)* 99.8%Fe	7.87	150	210	200	40
Fe35Mn, obtained by powder metallurgy and thermomechanically treated *(Hermawan, 2008)* Fe, 35.5% Mn, 0.04 C	-	235	550	-	32
FeMnPd, Cast and heat treated *(Schinhammer, 2009)* Fe, 10.2%Mn, 0.92%Pd, 0.12%C	-	850	1450	-	11
Magnesium, Heat treated condition *(ASM, 1998b)* 99.98% Mg	1.74	90	160	45	3
WE43, Heat treated condition *(ASTM, 2001)* Mg,3.7-4.3%Y,2.4-4.4%Nd,0.4 1%Zr	1.84	170	220	44	2
MgZnMnCa alloy, Cast condition *(Zhang, 2008)* Mg, 0.5%Ca, 2.0%Zn, 1.2%Mn	-	70	190	-	9
MgCa alloy, Extruded condition *(Li, 2008)* Mg, 1%Ca	1.74	140	240	45	11

* The values were used for comparative purposes, not being a metallic biomaterial that degrades in the human body. The values presented represent the minimum requirements imposed by the ASTM standard.

Materials in the composite category include biodegradable glass obtained by the rapid solidification of metal melts from some biodegradable metal alloys.

The idea of degradable biomaterials is moderately basic, since it includes their utilization in the execution of inserts that are processed in the human body after a specific timeframe, in which they play a particular practical part. Presently, there is a huge class of transitory inserts utilized in cardiovascular medical procedure, neurosurgery or muscular medical procedure, which require their extraction after culmination of the mending system of unhealthy tissue [45-46].

Biodegradable metals should offer satisfactory mechanical help to help the mending system during implantation. Nonetheless, it is hard to characterize a help period as close as conceivable to the clinical reality relying upon the different mediations that occur.

Fe or Mg-based combinations are two classes of metallic materials that have been proposed for making biodegradable inserts. On the off chance that on account of Fe-based composites just unadulterated Fe and amalgams from the Fe-Mn framework were examined, on account of Mg-based combinations a lot more extensive scope of Fe was proposed and explored. combinations: Mg-Al, Mg-RE and Mg-Ca. The point-by-point depiction of these magnesium compounds will be made in the accompanying passages [47, 48].

Concerning mechanical properties, it ought to be noticed that, albeit biodegradable metals have lower esteems than traditional metallic biomaterials (as per the information in Table 5.3), They are higher than those of biodegradable polymeric or composite biomaterials right now utilized in the execution of this class. of inserts, separately embeds for fixing breaks or cardiovascular stents [49].

5.3. Applications in medicine

Ahead of schedule in vivo and in vitro studies have obviously shown that the proposed biodegradable metal inserts have debased in the climate wherein they were presented, however the leftover serious issues are identified with the relationship of the biodegradation rate with the useful necessities forced on each sort of embed part [53, 54].

The orthopedic implant must resist mechanically until the callus is formed, and through resorption it must allow the gradual transfer of the load capacity to the bone, thus avoiding the occurrence of the "stress shielding" phenomenon and reducing the risk of post-implant fracture.

On account of a muscular embed, it would be great for resorption to happen through biodegradation and bone rebuilding. It should have osteoconductive properties, addressed by the capacity of a material to animate the development of bone tissue from the host issue that remains to be worked out embed and the arrangement of new, follower bone on the outer layer of the material (bone attachment/osseointegration). Thus, no other tissues are identified between the surface of the material and the newly formed bone tissue. A foreign object once implanted in the bone tissue will be covered by a film of blood that will later clot. Within about a week, the clotted blood layer will be replaced with granulation tissue. In the case of classical non-absorbable materials, this granulation tissue evolves towards the formation of a fibrous scar tissue. This process of fibrous encapsulation of the implant represents the so-called foreign body reaction, triggered by the action of macrophages. They initially try to phagocytose the foreign material, and later by failing this process, they will release cytokines that will induce the attraction and activation of fibroblasts that will encapsulate the object. The ideal resorbable material will divert this process to reconstituting the perimeter bone tissue instead of the fibrous encapsulation. Thus, the granulation tissue will be transformed into bone tissue by stimulating the activity of osteoblasts and osteoclasts (bone remodeling) and this newly formed bone tissue will adhere to the implant surface (osteointegration) [55,56].

Thus, an ideal resorbable implant can be characterized by:

1. Manufacturing material:

a. Biodegradable;

b. Biocompatible;

c. Bioactive (the ability of the interaction between the implant and the adjacent tissue). In the case of orthopedic implants to have the ability to be osteoconductive and osseointegration.

2. Mechanical properties, these must be maintained until the implant fulfills its purpose. For this, the following must be evaluated:

a. The elastic modulus - ideal would be to be as close as possible to the human bone. The differences between them lead to the effect of "stress shielding";

b. Tensile and compressive strength - preferably as close as possible to human bone values [6];

c. Fracture resistance is the ability of a material to withstand the spread of pre-existing cracks. This needs to be as large as possible to avoid rupture of the implant.

In conclusion, resorbable materials can give us the advantage of avoiding new ablation surgery, thus reducing the risks and stress to which the patient is subjected and reducing costs in the medical system. At the same time, by resorption they reduce the bone / implant contact stress and allow the transfer of the load capacity gradually from the implant to the bone, thus decreasing the risk of post-implant fracture, as it happens in the case of classic, non-absorbable implants [57].

Magnesium (Mg) as a chemical element was identified in 1808 by the British chemist Humphrey Davy and was isolated from Michael Faraday in 1852 by electrolysis of magnesium chloride, $MgCl_2$. As a resorbable medical device it was first used by Edward C. Huse in 1878 for ligating blood vessels. A disadvantage of these wires was their rigidity, the wires breaking at traction and the knots were difficult to make. To eliminate these disadvantages, an attempt was made to increase the ductility of magnesium by vacuum purification or alloying with gold

(Au) or silver (Ag), these experiments being conducted by Seelig [17] but the results were unsatisfactory [58, 59].

In 1917 EW Andrews used magnesium metal hemostatic clips, his idea being improved and patented in 1986 by Richard Jorgensen. The applications of magnesium in cardiology have progressed since 1998 through tests with biodegradable metal stents performed by the team led by Heublein.

Regarding the use of magnesium as a material for osteosynthesis, it was proposed by Payr in 1900, describing a nail made of magnesium used as a centromedullary implant to stabilize fractures or to cure pseudarthrosis. In 1906 Lambotte first used a resorbable metal implant made of magnesium, used to treat pseudarthrosis of a fracture of both calf bones in a 17-year-old man. It had been operated on unsuccessfully using conventional metal implants.

Lambotte [2] used a magnesium plate fixed with 6 steel screws, but had to remove it due to pain and inflammatory phenomena. He concluded that the mentioned phenomena were due to a severe and rapid degradation of the plate due to an electrochemical reaction generated by the contact between magnesium and steel, still not recommending the association of the magnesium implant with implants from other metals. He continued to test magnesium implants on rabbits and dogs, noting that resorption occurs completely between 7 and 10 months, by total degradation and pain is almost nil. Lambotte decided to use Mg implants in children, as they show faster bone healing. The first operation was performed on a 7-year-old child diagnosed with a supracondylar humerus fracture, followed by an 8-year-old boy with a humeral shaft fracture. They were operated using magnesium plates fixed with screws also made of magnesium (to prevent electrolytic corrosion) [60-63].

The results were positive, with magnesium fully resorbed at 1 year, and both fractures consolidating. In total, 4 children aged between 7 and 10 years were operated, the only reported complication being the accumulation of gas (hydrogen) in the periimplant soft tissues.

In 1913 Hey Groves used a centromedullary magnesium rod noting the rapid formation of the callus but also the equally rapid degradation of the rod. Verbrugge in 1934 improved the quality of magnesium implants by alloying them with aluminum (Al), obtaining an Mg-8Al alloy. He observed that the degradation was slower than in the case of pure magnesium, the resorption was total, and the hydrogen released did not produce any noticeable side effects. McBride obtained similar results by using an Mg-4Al-0.3Mn alloy, suggesting that magnesium can be used as an osteosynthetic material if the mechanical stresses to which it is subjected are reduced [64].

Table 5.4. Applications of magnesium [3, 65].

Author	Year	Magnesium	Alloy	Application
Huse	1878	Pure Mg	Wires for binding	Human model
Payr	1892-1905	Pure Mg	Plates, wires, tubes, anchors	Human model
Lambotte	1906-1932	Pure Mg	Plates, screws, pins	Human model, rabbits, dogs
Andrews	1917	Mg-Al alloy, Mg-Cd, Mg-Zn	Wires, connecting clamps	Dogs
Seelig	1924	Pure Mg distilled in vacuum	Wires, strips	Rabbits
Verbugge	1933-1937	Mg-Al6-Zn3-Mn0.2 alloy	Wires, plates, screws	Human model, dogs
McBride	1938	Mg-Mn3 alloy, Mg-Al4-Mn0.3	Wires, plates, screws	Rabbits, dogs
Stone	1951	Mg pure, aliaj Mg-Al2 alloy	Aneurysm binding wires	Dogs
Wexler	1980	Aliaj Mg-Al2	Intravascular threads	Mouse
Hussi	1981	Mg pure	Wires for hemangioma treatment	Mouse, rabbits
Wilflingseder	1981	Mg pure	Wires for hemangioma treatment	Human model

Despite all these promising results, research on the use of magnesium as an osteosynthetic material has stagnated for over 50 years, until the beginning of the 21st century.

Magnesium is currently a growing research topic, with a multitude of in vitro or in vivo studies testing various magnesium alloys for use as osteosynthesis materials. The use of magnesium, in various forms, in medical practice is also set out in the table 5.4.

Advantages of magnesium:

Magnesium has some physicochemical properties that recommend it as an excellent material for the manufacture of bioresorbable orthopedic implants. Thus, magnesium is a natural element of the human body, the magnesium implant having the ability to be completely resorbed without inducing local or systemic toxic effects. The amount of magnesium that the body needs daily is 240-420 mg / day, which is 20 to 50 times higher than needed for other metals, such as iron, 8-18 mg / day [66].

Magnesium has mechanical qualities that recommend it as a load-resistant implant, unlike polymers. It has a tensile strength of 160 - 250 MPa, while implants made of polymers have values of 16 - 69 MPa.

At the same time, it presents values of the elastic modulus of 41 - 45 GPa close to those of the human bone 10–40 Gpa, thus reducing the risk of stress shielding effect. This is manifested by the reduction of thickness and bone mass under contact with a metal implant (usually), due to the decrease of the moment of bending, torsion or compression forces, this being generated by a non-correlation between the elastic modulus of the implant and the bone. This risk is increased when using titanium or steel implants, which have increased values of elastic modulus, 110-117 GPa and 193 GPa respectively [67].

Another important advantage of magnesium, mentioned in the literature, is related to the stimulation of bone formation in the fracture focus, which integrates into the adjacent bone and allows a complete callus of the focus. It also has a beneficial effect on bone quality by increasing its strength.

Disadvantages of magnesium:

Despite all these qualities mentioned above, there are some potential negative aspects of magnesium in its use as a biomaterial for orthopedic implants. Elucidation and improvement of these disadvantages required the development of new processing techniques, in vitro tests and in vivo tests. The main suspicions were mainly related to: Low mechanical strength - although pure magnesium has values of elastic modulus close to those of bone, with a beneficial effect on the phenomenon of stress shielding, its overall mechanical strength is still low, predisposing to implant degradation. This is most likely when used in overloaded areas, such as the lumbar area where the forces to which the implant is subjected can exceed 3500 N [68].

Increased corrosion rate - this was one of the main problems encountered in using pure Mg as a biodegradable implant, the corrosion rate being faster than the rate of bone formation. Thus, the implant loses its mechanical qualities before the bone can take over the load.

Rapid elimination of magnesium in the body - due to the high rate of corrosion has raised the problem of rapid and excessive discharge of magnesium ions in the body, which can influence the serum values of magnesium with possible negative effects on systems and devices of the human body.

Release of hydrogen ions (H_2) - during the degradation process of magnesium hydrogen is released in the form of gas, which leads to the accumulation of gas bubbles in the tissues adjacent to the implant. This phenomenon has raised the suspicion of inducing tissue necrosis and late healing of tissues adjacent to gas bubbles. At the same time, the problem of gas

Advanced Metallic Biomaterials
Materials Research Foundations **118** (2022)

Materials Research Forum LLC
https://doi.org/10.21741/9781644901779

embolization arose, with the effect of vascular obstruction, producing areas of infarction in the vital organs (heart, lung) [69].

Hemolytic effect - pure magnesium has a slightly higher hemolysis rate, the values recorded varying between 25% and 75%. According to ISO 10993-5: 2009, a reduction in cell viability of over 30% is considered to be a cytotoxic effect [70].

Other applications of magnesium alloys:

- ❖ Magnesium, like aluminum, is durable and lightweight, so it is used in the construction of high-volume vehicles;

- ❖ Combined in alloys, it is essential in the construction of aircraft and missiles;

- ❖ Allied with aluminum, it improves its mechanical, processing and welding characteristics;

- ❖ Reducing agent for the production of pure uranium and other metals from their salts;

- ❖ Its hydroxide is used in "magnesium milk", magnesium chloride and sulfate in Epsom salts, and its citrates in medicine;

- ❖ Magnesium carbonate powder is used by athletes - gymnasts, weightlifters;

- ❖ Magnesium stearate is a white powder with lubricating properties. In the pharmaceutical industry it is used in the manufacturing process of tablets;

- ❖ It is also used in pyrotechnics and in the manufacture of incendiary bombs [70].

References

[1]   Yu, Y., Lu, H.; Sun, J. Long-term in vivo evolution of high purity Mg screw degradation — local and systemic effects of Mg degradation products. *Acta Biomater* **2018**, *71*, 215–224. https://doi.org/10.1016/j.actbio.2018.02.023

[2]   Ding W. Opportunities and challenges for the biodegradable magnesium alloys as next-generation biomaterials. *Regen Biomater* **2016**, *3(2)*, 79–76. https://doi.org/10.1093/rb/rbw003

[3]   Istrate, B. Materiale metalice biodegradabile pe bază de magneziu. *Ed. PIM*, Iasi, **2020**, ISBN: 978-606-13-5704-8.

[4]   Pogorielov, M.; Husak, E.; Solodivnik, A.; Zhdanov, S. Magnesium-based biodegradable alloys: degradation, application, and alloying elements. *Interv Med Appl Sci* **2017**, *9(1)*, 27–38. https://doi.org/10.1556/1646.9.2017.1.04

[5]   Tian, P.; Liu, X. Surface modification of biodegradable magnesium and its alloys for biomedical applications. *Regen Biomater* **2015**, *2(2)*, 135–151. https://doi.org/10.1093/rb/rbu013

[6]   Lukyanova, E.; Anisimova, N.; Martynenkoa, N.; Kiselevsky, M.; Dobatkina, S.; Estrin, Yu. Features of in vitro and in vivo behaviour of magnesium alloy WE43. *Mater Lett* **2018**, *215*, 308–311. https://doi.org/10.1016/j.matlet.2017.12.125

[7]   Wang, H.X.; Guan, S.K.; Wang, X.; Ren, C.X.; Wang, L.G. In vitro degradation and mechanical integrity of Mg–Zn–Ca alloy coated with Ca-deficient hydroxyapatite by the pulse electrodeposition process. *Acta Biomater* **2010**, *6(5)*, 1743–1748. https://doi.org/10.1016/j.actbio.2009.12.009

[8]   Gao, Y.; Wang, L.; Gu, X.; Chu, Z.; Guo, M.; Fan, Y. A quantitative study on magnesium alloy stent biodegradation. *J Biomech* **2018**, *74*, 98–105. https://doi.org/10.1016/j.jbiomech.2018.04.027

[9]   Atrens, A.; Song, G.L.; Liu, M.; Shi, Z.; Cao, F.; Dargusch, M.S. Review of recent developments in the field of magnesium corrosion. *Adv Eng Mater* 2015, *17(4)*, 400–453. https://doi.org/10.1002/adem.201400434

[10]   Hakimi, O.; Ventura, Y.; Goldman, J.; Vago, R.; Aghion, E. Porous biodegradable EW62 medical implants resist tumor cell growth. *Mater Sci Eng C Mater Biol Appl* **2016**, *61*, 516–525. https://doi.org/10.1016/j.msec.2015.12.043

[11]   Information on magnesium | Description, Properties, & Compounds. Available online: (accessed on September 10, 2021).

[12]   Gonzalez, J.; Hou, R.Q.; Nidadavolu, E.P.S; WillumeitRömer, R.; Feyerabend, F. Magnesium degradation under physiological conditions — best practice. *Bioact Mater* **2018**, *3(2)*, 174–185. https://doi.org/10.1016/j.bioactmat.2018.01.003

[13]   Wu, Y.; He, G.; Zhang, Y.; Liu, Y.; Li, M.; Wang, X.; Li, N.; Li, K.; Zheng, G.; Zheng, Y.; Yin, Q. Unique antitumor property of the Mg–Ca–Sr alloys with addition of Zn. *Sci Rep* **2016**, *6(1)*, 21736. https://doi.org/10.1038/srep21736

[14]   Walker, J.; Shadanbaz, S.; Woodfield, T.B.; Staiger, M.P.; Dias, G.J. Magnesium biomaterials for orthopedic application: a review from a biological perspective. *J Biomed Mater Res B Appl Biomater* **2014**, *102(6)*, 1316–1331. https://doi.org/10.1002/jbm.b.33113

[15]   Zhao, D.; Witte, F.; Lu, F.; Wang, J.; Li, J.; Qin, L. Current status on clinical applications of magnesium-based orthopaedic implants: a review from clinical

translational perspective. *Biomaterials* **2017**, *112*, 287–302.
https://doi.org/10.1016/j.biomaterials.2016.10.017

[16]   Staiger, M.P.; Pietak, A.M.; Huadmai, J.; Dias, G. Magnesium and its alloys as
orthopedic biomaterials: a review. *Biomaterials* **2006**, *27(9)*, 1728–1734.
https://doi.org/10.1016/j.biomaterials.2005.10.003

[17]   Shin, K.S.; Jung, H.C.; Bian, M.Z.; Nam, N.D.; Kim, N.J. Characterization of
biodegradable magnesium single crystals with various crystallographic orientations. *Eur
Cells Mater* **2013**, *26*, 4.

[18]   Witte, F.; Kaese, V.; Haferkamp, H.; Switzer, E.; Meyer-Lindenberg, A.; Wirth, C.J.;
Windhagen, H. In vivo corrosion of four magnesium alloys and the associated bone
response. *Biomaterials* **2005**, *26(17)*, 3557–3563.
https://doi.org/10.1016/j.biomaterials.2004.09.049

[19]   Peuster, M.; Beerbaum, P.; Bach, F.W.; Hauser, H. Are resorbable implants about to
become a reality? *Cardiol Young* **2006**, *16(2)*, 107–116.
https://doi.org/10.1017/S1047951106000011

[20]   Yazdimamaghani, M.; Razavi, M.; Vashaee, D.; Moharamzadeh, K.; Boccaccini,
A.R.; Tayebi, L. Porous magnesium-based scaffolds for tissue engineering. *Mater Sci
Eng C Mater Biol Appl* **2017**, *71*, 1253–1266. https://doi.org/10.1016/j.msec.2016.11.027

[21]   Lambotte, A. L'utilisation du magnésium comme matériel perdu dans l'ostéosynthèse
[The use of magnesium as material for osteosynthesis]. *Bull Mem Soc Nat Chir* **1932**, *28*,
1325–1334.

[22]   Wang, J.; Jiang, H.; Bi, Y.; Sun, Je.; Chen, M.; Liu, D. Effects of gas produced by
degradation of Mg–Zn–Zr alloy on cancellous bone tissue. *Mater Sci Eng C Mater Biol
Appl* **2015**, *55*, 556–561. https://doi.org/10.1016/j.msec.2015.05.082

[23]   Lee, J.W.; Han, H.S.; Han, K.J.; Park, J.; Jeon, H.; Ok, M.R.; Seok, H.K.; Ahn, J.P.;
Lee, K.E.; Lee, D.H.; Yang, S.J.; Cho, S.Y.; Cha, P.R.; Kwon, H.; Nam, T.H.; Han, J.H.;
Rho, H.J.; Lee, K.S.; Kim, Y.C.; Mantovani, D. Long-term clinical study and multiscale
analysis of in vivo biodegradation mechanism of Mg alloy. *Proc Natl Acad Sci USA*
2016, *113(3)*, 716–721. https://doi.org/10.1073/pnas.1518238113

[24]   Han, J.; Wan, P.; Ge, Y.; Fan, X.; Tan, L.; Li, J.; Yang, K. Tailoring the degradation
and biological response of a magnesium–strontium alloy for potential bone substitute
application. *Mater Sci Eng C Mater Biol Appl* **2016**, *58*, 799–811.
https://doi.org/10.1016/j.msec.2015.09.057

[25]   Verbrugge, J. Le matériel métallique résorbable en chirurgie osseuse [Resorbable metallic material in bone surgery]. *Presse Med 1934*; *23*, 460–465.

[26]   Onuma, Y.; Ormiston, J.; Serruys, P.W. Bioresorbable scaffold technologies. *Circ J* **2011**, *75(3)*, 509–520. https://doi.org/10.1253/circj.CJ-10-1135

[27]   Haude, M.; Ince, H.; Tölg, R.; Lemos, P.A.; von Birgelen, C.; Christiansen, E.H.; Wijns, W.; Neumann, F.J.; Eeckhout, E.; Garcia-Garcia, H.M.; Waksman, R. Sustained safety and performance of the second-generation drug-eluting absorbable metal scaffold (DREAMS 2G) in patients with de novo coronary lesions: 3-year clinical results and angiographic findings of the BIOSOLVE-II first-in-man trial. *EuroIntervention 2019*. https://doi.org/10.4244/EIJ-D-18-01000

[28]   Windhagen, H.; Radtke, K.; Weizbauer, A.; Diekmann, J.; Noll, Y.; Kreimeyer, U.; Schavan, R.; Stukenborg-Colsman, C.; Waizy, H. Biodegradable magnesium-based screw clinically equivalent to titanium screw in hallux valgus surgery: short term results of the first prospective, randomized, controlled clinical pilot study. *Biomed Eng Online* **2013**, *12(1)*, 62. https://doi.org/10.1186/1475-925X-12-62

[29]   Zhang, B.P.; Wang, Y.; Geng, L. Research on Mg–Zn–Ca alloy as degradable biomaterial. In: Biomaterials — physics and chemistry. *InTech* **2011**. https://doi.org/10.5772/23929

[30]   Liu C., Ren Z., Xu Y., Pang S., Zhao X., Zhao Y. Biodegradable magnesium alloys developed as bone repair materials: a review. *Scanning 2018*, **2018**, 9216314. https://doi.org/10.1155/2018/9216314

[31]   Thomann, M.; Krause, C.; Bormann, D.; von der Höh, N.; Windhagen, H.; Meyer-Lindenberg, A. Comparison of the resorbable magnesium. Alloys LAE442 und MgCa0.8 concerning their mechanical properties, their progress of degradation and the bone-implant-contact after 12 months implantation duration in a rabbit model. *Materwiss Werksttech* **2009**, *40(1–2)*, 82–87. https://doi.org/10.1002/mawe.200800412

[32]   Farraro, K.F.; Kim, K.E.; Woo, S.L.Y.,; Flowers, J.R.; McCullough, M.B. Revolutionizing orthopaedic biomaterials: the potential of biodegradable and bioresorbable magnesium based materials for functional tissue engineering. *J Biomech* **2014**, *47(9)*, 1979–1986. https://doi.org/10.1016/j.jbiomech.2013.12.003

[33]   Sasikumar, Y.; Kumar, A.M.; Babu, R.S.; Rahman, M.M.; Samyn, L.M.; de Barros, A.L.F. Biocompatible hydrophilic brushite coatings on AZX310 and AM50 alloys for orthopaedic implants. *J Mater Sci Mater Med* **2018**, *29(8)*, 123. https://doi.org/10.1007/s10856-018-6131-8

[34]   Istrate, B.; Rau, J.V.; Munteanu, C.; Antoniac, I.V.; Saceleanu, V. Properties and in vitro assessment of ZrO$_2$-based coatings obtained by atmospheric plasma jet spraying on biodegradable Mg-Ca and Mg-Ca-Zr alloys. *CERAMICS INTERNATIONAL* **2020**, *46(10)*, 15897-15906. https://doi.org/10.1016/j.ceramint.2020.03.138

[35]   Nene, S.S.; Kashyap, B.P.; Prabhu, N.; Estrin, Y.; Al-Samman, T. Biocorrosion and biodegradation behavior of ultralight Mg–4Li–1Ca (LC41) alloy in simulated body fluid for degradable implant applications. *J Mater Sci* **2015**, *50(8)*, 3041–3050. https://doi.org/10.1007/s10853-015-8846-y

[36]   Eddy Jai Poinern, G.; Brundavanam, S.; Fawcett, D. Biomedical magnesium alloys: a review of material properties, surface modifications and potential as a biodegradable orthopaedic implant. *Am J Biomed Eng* **2013**, *2(6)*, 218–240. https://doi.org/10.5923/j.ajbe.20120206.02

[37]   Castellani, C.; Lindtner, R.A.; Hausbrandt, P.; Tschegg, E.; Stanzl-Tschegg, S.E.; Zanoni, G.; Beck, S.; Weinberg, A.M. Bone implant interface strength and osseointegration: biodegradable magnesium alloy versus standard titanium control. *Acta Biomater* **2011**, *7(1)*, 432–440. https://doi.org/10.1016/j.actbio.2010.08.020

[38]   Wang, H.; Estrin, Y.; Zúberová, Z. Bio-corrosion of a magnesium alloy with different processing histories. *Mater Letters* **2008**, *62(16)*, 2476–2479. https://doi.org/10.1016/j.matlet.2007.12.052

[39]   Wu, Z.; Curtin, W.A. The origins of high hardening and low ductility in magnesium. *Nature* **2015**, *526(7571)*, 62–67. https://doi.org/10.1038/nature15364

[40]   Feyerabend, F.; Fischer, J.; Holtz, J.; Witte, F.; Willumeit, R.; Drucker, H.; Vogt, C.; Hort, N. Evaluation of short-term effects of rare earth and other elements used in magnesium alloys on primary cells and cell lines. *Acta Biomater* **2010**, *6(5)*, 1834–1842. https://doi.org/10.1016/j.actbio.2009.09.024

[41]   Harandi, S.E.; Mirshahi, M.; Koleini, S.; Idris, M.H.; Jafari, H.; Kadir, M.R.A. Effect of calcium content on the microstructure, hardness and in-vitro corrosion behavior of biodegradable Mg–Ca binary alloy. *Mater Res* **2013**, *16(1)*, 11–18. https://doi.org/10.1590/S1516-14392012005000151

[42]   Mareci, D.; Bolat, G.; Izquierdo, J.; Crimu, C.; Munteanu, C.; Antoniac, I.; Souto, R.M. Electrochemical characteristics of bioresorbable binary MgCa alloys in Ringer's solution: revealing the impact of local pH distributions during in-vitro dissolution. *Mater Sci Eng C Mater Biol Appl* **2016**, *60*, 402–410. https://doi.org/10.1016/j.msec.2015.11.069

[43]   Makkar, P.; Sarkar, S.K.; Padalhin, A.R.; Moon, B.G.; Lee, Y.S.; Lee, B.T. In vitro and in vivo assessment of biomedical Mg–Ca alloys for bone implant. *J Appl Biomater Funct Mater* **2018**, *16(3)*, 126–136. https://doi.org/10.1177/2280800017750359

[44]   Kannan, M.B.; Raman, R.K. In vitro degradation and mechanical integrity of calcium containing magnesium alloy in modified simulated body fluid. *Biomaterials* **2008**, *29*, 2306–2314. https://doi.org/10.1016/j.biomaterials.2008.02.003

[45]   Kirkpatrick, C.J.; Alves, A.; Köhler, H.; Kriegsmann, J.; Bittinger, F.; Otto, M.; Williams, D.F.; Eloy, R. Biomaterial induced sarcoma: a novel model to study preneoplastic change. *Am J Pathol* **2000**, *156(4)*, 1455–1467. https://doi.org/10.1016/S0002-9440(10)65014-6

[46]   Agha, N.A.; Liu, Z.; Feyerabend, F.; Willumeit Römer, R.; Gasharova, B.; Heidrich, S.; Mihailova, B. The effect of osteoblasts on the surface oxidation processes of biodegradable Mg and Mg-Ag alloys studied by synchrotron IR microspectroscopy. *Mater Sci Eng C Mater Biol Appl* **2018**, *91*, 659–668. https://doi.org/10.1016/j.msec.2018.06.001

[47]   Zhao, N.; Zhu, D. Endothelial responses of magnesium and other alloying elements in magnesium-based stent materials. *Metallomics* **2015**, *7(1)*, 118–128. https://doi.org/10.1039/C4MT00244J

[48]   Chen, Y.M.; Xiao, M.; Zhao, H.; Yang, B.C. On the antitumor properties of biomedical magnesium metal. *J Mater Chem B* **2015**, *3(5)*, 849–858. https://doi.org/10.1039/C4TB01421A

[49]   Kavalar, R.; Fokter, S.K.; Lamovec, J. Total hip arthroplasty-related osteogenic osteosarcoma: case report and review of the literature. *Eur J Med Res* **2016**, *21(1)*, 8. https://doi.org/10.1186/s40001-016-0203-3

[50]   Witte, F.; Hort, N.; Vogt, C.; Cohen, S.; Kainer, K.U.; Willumeit, R.; Feyerabend, F. Degradable biomaterials based on magnesium corrosion. *Curr Opin Solid State Mater Sci* **2008**, *12(5–6)*, 63–72. https://doi.org/10.1016/j.cossms.2009.04.001

[51]   Sanz-Herrera, J.A.; Reina-Romo, E.; Boccaccini, A.R. In silico design of magnesium implants: macroscopic modeling. *J Mech Behav Biomed Mater* **2018**, *79*, 181–188. https://doi.org/10.1016/j.jmbbm.2017.12.016

[52]   Kirkland, N.T. Magnesium biomaterials: past, present and future. *Corros Eng Sci Technol* **2012**, *47(5)*, 322–328. https://doi.org/10.1179/1743278212Y.0000000034

[53]   Song, G.; Atrens, A. Understanding magnesium corrosion: a framework for improved alloy performance. *Adv Eng Mater* **2003**, *5*, 837–858. https://doi.org/10.1002/adem.200310405

[54]   Nidadavolu, E.P.S.; Feyerabend, F.; Ebel, T.; Willumeit-Römer, R.; Dahms, M. On the determination of magnesium degradation rates under physiological conditions. *Materials (Basel)* **2016**, *9(8)*, E627. https://doi.org/10.3390/ma9080627

[55]   Jiang W., Tian Q., Vuong T., Shashaty M., Gopez C., Sanders T., Liu H. Comparison study on four biodegradable polymer coatings for controlling magnesium degradation and human endothelial cell adhesion and spreading. *ACS Biomater Sci Eng* **2017**, *3(6)*, 936–950. https://doi.org/10.1021/acsbiomaterials.7b00215

[56]   Brooks, E.K.; Ehrensberger, M. Bio-corrosion of magnesium alloys for orthopaedic applications. *J Funct Biomater* **2017**, *8(3)*, 38. https://doi.org/10.3390/jfb8030038

[57]   Neacsu, P.; Staras, A.I.; Voicu, S.I.; Ionascu, I.; Soare, T.; Uzun, S.; Cojocaru, V.D.; Pandele, A.M.; Croitoru, S.M.; Miculescu, F.; Cotrut, C.M.; Dan, I.; Cimpean, A. Characterization and in vitro and in vivo assessment of a novel cellulose acetate-coated Mg-based alloy for orthopedic applications. **Materials (Basel)** *2017*, *10(7)*, 686. https://doi.org/10.3390/ma10070686

[58]   Bian, D.; Deng, J.; Li, N.; Chu, X.; Liu, Y.; Li, W.; Cai, H.; Xiu, P.; Zhang, Y.; Guan, Z.; Zheng, Y.; Kou, Y.; Jiang, B.; Chen, R. In vitro and in vivo studies on biomedical magnesium low alloying with elements gadolinium and zinc for orthopedic implant applications. *ACS Appl Mater Interfaces* **2018**, *10(5)*, 4394–4408. https://doi.org/10.1021/acsami.7b15498

[59]   Ralston, K.; Birbilis, N.; Davies, C. Revealing the relationship between grain size and corrosion rate of metals. *Scr Mater* **2010**, *63(12)*, 1201–1204. https://doi.org/10.1016/j.scriptamat.2010.08.035

[60]   Dobatkin, S.V.; Lukyanova, E.A.; Martynenko, N.S.; Anisimova, N.Yu.; Kiselevskiy, M.V.; Gorshenkov, M.V.; Yurchenko, N.Yu.; Raab, G.I.; Yusupov, V.S.; Birbilis, N.; Salishchev, G.A.; Estrin, Yu.Z. Strength, corrosion resistance, and biocompatibility of ultrafine-grained Mg alloys after different modes of severe plastic deformation. *IOP Conference Series: Materials Science and Engineering* **2017**, *194*, 012004. https://doi.org/10.1088/1757-899X/194/1/012004

[61]   Zhou, Y.L.; Li, Y.; Luo, D.M.; Ding, Y.; Hodgson, P. Microstructures, mechanical and corrosion properties and biocompatibility of as extruded Mg–Mn–Zn–Nd alloys for biomedical applications. *Mater Sci Eng C Mater Biol Appl* **2015**, *49*, 93–100. https://doi.org/10.1016/j.msec.2014.12.057

[62]   Bian, D.; Zhou, W.; Liu, Y.; Li, N.; Zheng, Y.; Sun, Z. Fatigue behaviors of HP–Mg, Mg–Ca and Mg–Zn–Ca biodegradable metals in air and simulated body fluid. *Acta Biomater* **2016**, *41*, 351–360. https://doi.org/10.1016/j.actbio.2016.05.031

[63]   Jiang, W.; Cipriano, A.F.; Tian, Q.; Zhang, C.; Lopez, M.; Sallee, A.; Lin, A.; Cortez Alcaraz, M.C.; Wu, Y.; Zheng, Y.; Liu, H. In vitro evaluation of MgSr and MgCaSr alloys via direct culture with bone marrow derived mesenchymal stem cells. *Acta Biomater* **2018**, *72*, 407–423. https://doi.org/10.1016/j.actbio.2018.03.049

[64]   Kirkland, N.T.; Birbilis N.; Staiger, M.P. Assessing the corrosion of biodegradable magnesium implants: a critical review of current methodologies and their limitations. *Acta Biomater* **2012**, *8(3)*, 925–936. https://doi.org/10.1016/j.actbio.2011.11.014

[65]   Kirkland, N.T.; Lespagnol, J.; Birbilis, N.; Staiger, M.P. A survey of bio-corrosion rates of magnesium alloys. *Corros Sci* **2010**, *52(2)*, 287–291. https://doi.org/10.1016/j.corsci.2009.09.033

[66]   Cipriano, A.F.; Lin, J.; Miller, C.; Lin, A.; Cortez Alcaraz, M.C.; Soria, P.; Liu, H; Anodization of magnesium for biomedical applications — processing, characterization, degradation and cytocompatibility. *Acta Biomater* **2017**, *62*, 397–417. https://doi.org/10.1016/j.actbio.2017.08.017

[67]   Li, M.; Wang, W.; Zhu, Y.; Lu, Y.; Wan, P.; Yang, K.; Zhang, Y.; Mao, C. Molecular and cellular mechanisms for zoledronic acid-loaded magnesium-strontium alloys to inhibit giant cell tumors of bone. *Acta Biomater* **2018**, *77*, 365–379. https://doi.org/10.1016/j.actbio.2018.07.028

[68]   Istrate, B.; Mareci, D.; Munteanu, C.; Stanciu, S.; Crimu, C.I.; Trinca, L.C.; Kamel, E. In vitro electrochemical properties of biodegradable YSZ-coated MgCa alloy. *Environmental Engineering and Management Journal*, **2016**, *15(5)*, 955-963. https://doi.org/10.30638/eemj.2016.104

[69]   Sindilar, E.V.; Munteanu, C.; Pasca, S.A.; Mihai, I.; Henea, M.E.; Istrate, B. Long Term Evaluation of Biodegradation and Biocompatibility In-Vivo the Mg-0.5Ca-xZr Alloys in Rats. *Crystals*, **2019**, *11(1)*, 54. https://doi.org/10.3390/cryst11010054

[70]   Istrate, B.; Munteanu, C.; Chelariu, R.; Mihai, D.; Cimpoesu, R.; Tudose, F.S.V. Electrochemical Evaluation of Some Mg-Ca-Mn-Zr Biodegradable Alloys. *Revista de Chimie* **2019**, *70(9)*, 3435-3440, 2019. https://doi.org/10.37358/RC.19.9.7565

CHAPTER 6

Optimization of Metallic Biomaterials

6.1. Brief presentation of the biological environment

There are two relative quantitative aspects that distinguish the interactions of biomaterials with the biological environment and create the need for independent study of the host's and material responses [1-4]:

1. Specific requirements – the biological environment, especially the internal environment of living systems is very aggressive; is an environment with an intense and complex chemical activity, combined with a wide spectrum, variable according to some mechanical stresses combined.

2. Stable conditions – despite the aggressive aspects, the biological environment presents an extraordinary constancy of both physical conditions and composition. There are complex control systems that ensure this constancy. Consequently, deviations of the stable conditions due to the presence of the material can cause appropriate responses.

The forceful parts of the organic climate can be perceived on the off chance that we look at the contrasts between the interior and outside states of living frameworks. Remotely, we track down the natural parts of the encompassing scene. By far most of materials are inorganic and are to some extent or completely oxidized. While actual cycles are interrelated, there is an absence of dynamic control frameworks. The trade time constants are for some time, controlled by compound dissemination processes and composed by sources that give energy primarily through radiation, conduction and convection. Many atomic species are known, but there is a wide variety of chemical structure and content and limited possibilities for structural and compositional optimization.

Paradoxically, the inside climate seems, by all accounts, to be a framework wherein materials (particles and tissues) are generally natural in nature and are somewhat or completely diminished in size. The vast majority of changes are mediated as active control systems that require energy. In many cases, multiple parallel systems with different time constants with extended interactions on multiple systems, controlling a single transformation or process. Although there are a multitude of compounds and chemical structures, they are mainly made up of a few basic elements namely carbon, oxygen, hydrogen and nitrogen. The present elements are used to the maximum, and the structures have a great efficiency, giving the impression of an optimal design and arrangement [5-6].

Regardless of the views on the development and origin of biological systems, everyone is impressed by their complexity. They perform their functions by accidentally excluding, by design or by physico-chemical processes, materials that are not necessary, or that harm their function in individual processes. These phenomena act to exclude all materials that are sick or that are not part of the body. Moreover, the system interacts both locally and regionally or globally. Therefore, a constant aspect of the biological environment is that the introduction of a foreign material will cause a response of the host, which can have consequences at the local or system level [7].

It is hard to characterize the biological environment wherein a gadget or material plays out a specific capacity. The trouble emerges in the absence of itemized information about the in vivo conditions and about the nearby varieties that might happen in the cycles of arrangement and creation of homeostasis, totally essential for life [8,9].

There is likewise a specific equivocalness in characterizing the locale at the interface where the natural climate couples with a material. Inserts from detached locales of the body can associate with the remainder of the framework through the dissemination of particles and liquids, through blood flow and through the waste of lymph. In any event, characterizing the outright volume of material that speaks with an embed can be troublesome. Materials should be tried in vitro before implantation even in creatures. It is desirable to try to obtain in the laboratory the operational environment that the material will meet after implantation [10].

In the case of in vivo testing, this shall be carried out only under controlled physiological and biophysiological conditions. The biological environment is generally regarded as the sum of the conditions that an implanted material will encounter chronically or acutely, if it is the combination of biological and pericellular conditions. Combining these extrinsic and intrinsic effects of the environment with the overall requirements of the patient during the proposed period for implantation is called the history of the life of the implant, meaning the totality of the requirements that the biomaterial must meet in order to be successful in the application [11].

The thermal, mechanical, chemical and surface parameters are sufficient to describe in general the biological, violent environment that the implant encounters. The values differ slightly from patient to patient; the differences that exist influence too little the response of the host and the material. It is a pity that the technology for determining the functionality of the implant and of the biomaterial-biological environment interactions is poorly developed compared to that available to scientists in the biological field, who study the organs in situ.

In a specific application, the selection of materials and the design of the elements that incorporate them is called "appropriate expectations". This term does not take into account the changes that occur in the patient's life after implantation, but has the role of guiding the selection of technologies before implantation. It is better said that we are trying to design the most durable biomaterials and the best surfaces that best meet the requirements [12].

When performing the tests necessary for the evaluation of the interactions from the biomaterial-tissue interface, it is taken into account the objective and subjective factors that can influence the tissue response, but also that of the biomaterial, in order to make a correct interpretation of the results, depending on the material factors and specifically, those related to the surface [13].

As a rule, there are two transitional cycles normal to all implantation applications. In the first place, the embed might be sullied incidentally or because of assembling or taking care of cycles, during capacity or addition. It is normally expected that the outer layer of the embed is unadulterated and clean. Reality can be very surprising. Natural components might continue during fabricate or because of ill-advised taking care of. Oxidation or other destructive components might happen during the preoperative stages; materials can be gotten from bundling utilized for their capacity, and microorganisms can be moved from careful instruments [14, 15].

For this reason, experimental studies on biomaterial-tissue interaction should include surface characteristics of current implant samples selected from a batch manufactured for a particular study under conditions prior to surgical insertion.

Secondly, all implants must be sterilized and cleaned before use. Some of them can be delivered by the manufacturer in sterile packages with two coatings, while others must be sterilized in the laboratory or hospital before use. The most common forms of sterilization used in the practice of the implant are: cold solution, dry heat, moist heat, gas, gamma radiation. If the sterilization process is over, both the host's and the material's responses may be affected. Sterilizing an implant can make it sterile but not risk-free, thus altering the host's response. Therefore, in examining the response of the material or host to the implant, it is necessary to pay increased attention to the conditions of surface preparation [16].

The success of an implantation replacement depends a lot on the interface between the tissue and the synthetic material, where totally different effects can be achieved. While for implants that remain in contact with blood, hemocompatible behavior with minimal interaction is normally required, osteointegrated implants must exhibit a strong interaction in order to obtain a high adhesion force. Osseointegration can be influenced both by the structure and topography of the surface, as well as by its composition [17, 18].

6.2. Factors influencing the interaction of biomaterial-tissue

Important parameters for direct tissue-biomaterial contact are the biocompatibility of the material chosen for the implant, the macrostructure (shape), the microstructure (surface roughness, geometry of the elevations and recesses in the implant), the implantation surgical procedure, the direct tissue-implant contact after insertion (implantation), the time and mode of loading of the implant, which can lead to the production of relative movements of the implant in relation to the adjacent tissue and the implant support is. Recently, it has become apparent that the microstructure of the implant surface influences in a significant way the osteoblasts, but also the amount of tissue formed at the interface. Therefore, the biocompatibility of an implant is just one of the parameters that influence the tissue's response to the metal implant; also, the morphology of surfaces, of microscopic structures becomes important, because the main problem that limits the application and operation of metal implants is the lack of a viable anchoring of the implant in the tissue. On this basis, experimentally, there were conducted studies on implants with different macrostructures, which were used to obtain a better anchorage and a long-term stabilization of the implantation support [19-21].

One way to solve the problem of tissue anchoring at the surface of the implant is to use structured surfaces that allow cells to grow, such as porous surfaces. It is known that implants sandblasted or those covered with spherical particles allow the contact of the tissue and its growth in the interstices of the implant [22].

A second way to improve the surface structure of metal implants would be to produce controlled certain microstructures, a surface roughness of the order of a few micrometers. The literature specifies that the surface structure of a metal implant should allow a secondary fixation of the implant by penetrating the bone trabeculae into the surface microcavitations. Observing such surfaces of the implants under the scanning electron microscope, it was found a similarity with the surface of the bone after resorption. In the articles published in the 1980s it is specified that such resorbed bone surfaces have in many cases used as anchor points for the newly formed bone. After obtaining these results, it is expected that the surfaces of the implants that mimathe the bone structure after resorption will be more successful than others, being fixed to the surrounding bone. Therefore, it is of great importance to better describe, quantitatively and qualitatively, the surface of the bone and the implant. In the specialized literature is described in detail, the morphology of the surface of implants by three-dimensional methods, being quantified the height and distances between elevations, by using different quantitative testing methods [23-25].

The principle of direct contact between the tissue and the implant involves rigid mechanical fixation, because the growing bone is separated from the implant by the appearance of a soft layer of tissue, an interfacial membrane that can activate micro-joints at the interface. It is known that micro-motions can produce undesirable reactions, such as the destruction of the oxide layer on the surface of the implant, which can be followed by the appearance of the corrosion phenomenon and the rejection of the implant. The larger surface area of rough implants in contact with tissue also poses a higher risk of corrosion processes. In addition, rigid metal implants with a certain geometry, rigidly fixed in the tissue, can cause shielding, which can lead to bone resorption. The possibility of these unwanted reactions points out once again the importance of surface preparation, which includes the geometry of the implant, the arrangement and amplitude of the surface elevations and recesses [26].

The surface design of an embed has a huge impact, both in its obsession and in the power of attachment to the tissue. Since this power is characterized as a small amount of the squeezing power on a superficial level on which it is applied, the heap under genuine conditions can be the more noteworthy the bigger the space of the surface. From this basic thought, unmistakably a delicate surface of the embed, with a little space of contact with the tissue, presents a lower bond power than an organized surface [27-28].

Some authors who have studied the importance of roughness in the case of biomaterial-tissue interaction through the use of cylindrical implants have presented results that prove that, by increasing the roughness, the adhesion force increases. It has also been found that the tensile strength increases with the increase of the contact areas by making holes in the flat-fired cylinders. With the increase of the surface (by drilling) the resulting stretching voltage was calculated. In addition, the strength of adhesion depends on the implantation time. The bone needs a certain amount of time to fill the free space, by growing, between the implant and the surface cavities, causing a mechanical stress relief of the implant. From the point of view of mechanical strength, bone growth in surface cavities has a favorable influence. The shear voltage, which is generated by the functional loading of the implant to the biomaterial-tissue interface, is low because, similarly to the wear of a screw, the load on this interface, combined with the tension perpendicular to the inclined area, involves only a small part of the shear tension that actually acts on the inclined surface [29, 30].

Apart from the favorable influence on implant fixation, there is an important advantage in using surface topography structuring. The growing bone, which is subjected to compression tension, produces calcium, thus stimulating the growth of a bone formation. It is clear that the emergence of this phenomenon will intensify in proportion to the improvement in loading transfer [31-34].

Several authors found that the implanted cylindrical samples, with a polished surface ($Rt>1\mu m$) compared to those with a rough surface ($Rt=20\mu m$), showed a significantly weaker behavior in terms of fixation. In the case of polished samples, the bone trabecules were oriented perpendicular to the surface of the implant, while in the case of implants with the rough surface of the formed trabecules, concentrically arranged around it. In addition, a space of soft tissue was observed between the polished implants and the bone, while on the samples with a porous layer, a better contact was observed [35].

The biological environment in the human body is a chemically, mechanically and electrically active environment, and the interface between biomaterial and tissue is a place where a series of biochemical and biodynamic processes and reactions take place. Oxygen diffuses from the oxidized surface into the base metal, and metal ions can also diffuse over the surface. The interactions of biological molecules with the surface of the implant can cause transient or permanent changes in the conformation of these molecules and thereby, functional changes [36].

6.3. The influence of geometric factors on the interaction of biomaterial-tissue

Geometric factors are extremely important when considering the interactions that take place between the implant and infectious microorganisms. Initiation (planting or colonization) and propagation of infection is a competitive process between the invading bacteria and the host tissue. Therefore, the exposure of a bacterium to a tissue, and especially to vascular processes, is a critical factor in the success of the host's defense against infection. The tissue around an implant is relatively acellular in adulthood. Bacteria that can grow in the capsular membrane will encounter very little opposition until they enter the surrounding tissue. Aside from the general and specific immune response, the immediate response to infection is inflammation. The cells that mediate the response come from the blood. Reducing the angle of the tissue to the implant is of great importance, reducing the access of the infected tissue near the implant to neutrophils and macrophages [37, 38].

The existence of a "dead" space, a volume filled with fluid that does not contain cells, represents a risk that can occur due to the geometric shape of the implant. Medical practice tries to avoid this feature because the liquid can act as well as the culture of bacteria in vivo [39].

Even in the absence of an acellular membrane or a dead space, certain bacteria can produce their own geometric defense system by forming glycolax around them on the surface of the implant. Although porous materials do not represent a particular structural risk from the point of view of carcinogenesis, they may present a geometric risk when initiating and supporting the infection [40-41].

The clinical practice of removing the infected implant is based on geometric arguments rather than on their physiological consequences. Two other types of arguments can be proposed: one based on chemically mediated interactions between the implant and bacterial infection, and the other based on chemical effects on the host's cell populations attacking the infection. Both are controversial today. It is believed that if implants can provide metals necessary for metabolism by corroding their metal parts, it can be shown that they can also participate in the metabolism of bacteria. The metal implant has a similar effect to that of garden fertilizer. For example, a metal, iron, occupies a special place. The level of iron in the body changes rapidly in response to infection. It has been suggested that the interference of the indigenous iron source with the control system plays a role in the propagation of certain infections [42-43].

6.4. Possible reactions to the biomaterial-tissue interface

Albeit the interface between a biomaterial and the body is very perplexing, the results of tissues that join living cells into a biomaterial framework make things considerably more muddled. These items have three unmistakable interfaces to consider: between the body and the biomaterial, between the body and living cells, among cells and biomaterial. Every one of these interfaces presents potential issues identified with the drawn-out accomplishment of the embed. Researchers have long sought to understand and control the events that take place at the level of these interfaces. Unele books considered classics, published 10 or 20 years ago, are still topical. As more is learned about the beneficial and damaging clinical events taking place at the tissue-biomaterial interface, new ideas emerge in the design of the biomaterial surface to control the functions of cells and tissues and to produce new biomaterials that can integrate with the tissues of the body. The most straightforward type of cooperation among inserts and the organic climate is the exchange of material to the material-tissue interface, without a response. On the off chance that the substance, particles or liquid, relocates from tissue into biomaterial, it brings about an extremely thick material, swell, because of the protection of volume. Even in the absence of fluid takeover, the biomaterial can absorb certain components or dissolved substances from the surrounding fluids. If the fluid enters the tissue or if a component of the biomaterial dissolves in the tissue fluids, the porosity of the resulting material is due to the filtration process. Both effects can have profound consequences for the behavior of the materials, despite the absence of mechanical stresses applied from the outside and the obvious change in shape. Swelling and filtration are the result of the diffusion process. Discussing in strict terms, in a biomaterial-tissue system there are no surfaces, only interfaces. The solid-liquid interface can affect the dissolved elements in the surrounding fluid. The chemical effects depend on the details of the chemical and ionic distribution on the surface. These elements may have negative effects on the chosen biomaterial or may deliberately induce the effects necessary to mediate the cellular response [44-45].

Metals, in touch with the natural creature, give complex impacts, delivering a progression of organic responses, contingent upon the fixation, the openness time, and so on as indicated by the method of natural collaboration, the metals are separated into:

- Metal elements necessary in very low concentrations, for the living organism, called essential elements, among which are reminiscent: cobalt, iron, manganese, zinc, magnesium, sodium, potassium, etc.;

- Elements that produce toxic reactions for the body, if present in higher concentrations, such as arsenic, cobalt, nickel, etc.; the cytotoxic effect has been demonstrated by the system of cell cultures;

- Metals with allergic potential as nickel, cobalt and chromium are considered as strongly allergic elements;

- Metals and some metal compounds have a carcinogenic effect by forming free radicals in contact with the biological environment [46-48].

6.4.1. Toxicity of metals

In biological environments, in low concentrations metals do not cause toxic or allergic effects.

Only in high concentrations, and especially in the form of metallic compounds (oxides, salts, etc.) biological reactions harmful to the body are produced. The behavior of the most common metal elements in biological environments is presented below:

Poisoning with metallic aluminum is found in practice quite often. The main route of entry into the body is respiratory, by inhalation. Aluminum compounds are slowly absorbed, the target organ being the brain.

Aluminum negatively influences bone metabolism by inhibiting the process of phosphorylation and ATP synthesis, thus reducing the cellular energy reserve. When the amount of aluminum is quite large in the bone tissue, its mineralization is altered, thus causing pathological fractures. The major problem of Al is that it accumulates in the cells of the nervous system [49, 50].

Beryl can enter the body in different ways: cutaneous, respiratory or digestive. It has been experimentally proven that berylium and its derivatives are carcinogenetic, and even a few cases of pneumoconiosis and beryliosis have been reported in dental technicians.

The respiratory tract is the main access route for cadmium and its derivatives in the body, especially due to smoking and occupational exposures. The target human organ is the lung, after inhalation of oxides and sulphites, and the most common diseases are bronchitis, bronchopneumonia and pulmonary edema [52].

The absorption of cobalt occurs through digestive, cutaneous and pulmonary pathways. Cobalt is stored in the liver, kidneys and pancreas, although 20% of cobalt is eliminated from the body within a few days, mostly by the urinary route. Cobalt absorption is the main cause of cases of pulmonary fibrosis induced by heavy metals. Cobalt holds the 3rd place, after nickel and chromium, among allergenic metals.

Chromium additionally has a place with the gathering of components fundamental forever. Its poisonousness relies upon the oxidation state where it is found. Hexavalent chromium is more poisonous than trivalent chromium, being viewed as an amazing mutagenic and cancer-causing specialist. Trivalent mixtures don't enter through the skin or cell layers and tie to stable protein buildings. Unexpectedly, hexavalent compounds have a higher oxidation force of natural atoms, effectively going through cell layers and being diminished to the more steady trivalent structure, which enters the core and instigates transformations through cooperations with DNA. Chromium is absorbed mainly through the digestive tract and, secondaryly, through the skin and lung. Chromium can accumulate in the liver, uterus, kidneys and bones, the accumulation being greater for the hexavalent form. Chromium has great allergenic potential [53].

Copper is absorbed through the digestive tract and airways. Copper is quickly excreted through the bile. From a clinical point of view, copper poisoning can lead to: granulomatous fibrosis in the lung, micronodular cirrhosis or hemangiosarcoma in the liver, cellular necrosis in the kidneys, cellular lysis in the epithelium of the mucous membranes [54].

Iron is an essential element, ubiquitous in the body. He has a central role in the transport of oxygen. Iron is toxic only by exposure to very high levels.

Manganese has no toxic effect except in extreme cases of occupational exposure. It is an essential element with an important role in the activation of some enzymatic systems.

Molybdenum is also considered an essential element of life, people needing a daily intake of 0.1 mg Mo. Pulmonary edema and gout symptoms have been observed in workers with occupational exposures by prolonged inhalation of metallic molybdenum particles.

Nickel is one of the most concentrated on components comparable to the impacts on the human body. It is universal in the climate in which we live and is consumed on stomach related, respiratory, cutaneous or metal inserts, by inward breath, implantation, ingestion, intraperitoneal, intramuscular or intravenous infusions. The measure of nickel in the blood and pee is a reliable pointer of inebriation with this metal. The objective organ is the lung. Nickel is viewed as the most sharpening metal for delicate tissues. It prompts hypersensitive contact dermatitis, causing more unfavorably susceptible responses than any remaining metals consolidated [55].

6.4.2. Ion release

The lesions caused by the use of metal alloys as biomaterials are mainly due to the release of ions, ions resulting from the corrosion of these alloys. These released ions are mainly the following:

• nickel, cobalt and chromium – for any application;

• beryl, cadmium, palladium, silver and copper – for dental alloys;

• titanium – for metal alloys used in the execution of orthopedic and dental implants.

Laborious investigations on the effects and reactions produced by the release of ions are rare and no statistical and epidemiological studies have not been made, although the releases of ions by metals such as nickel, chromium and cobalt are well known. The ions released by the prostheses and metal implants, as a result of the electrochemical corrosion process, are the main source of reactions with the biological environment [56, 57].

6.4.3. Tissue reactions

In the vicinity of implants or prostheses with biological tissue, tissue reactions may occur. These reactions between biomaterials and the biological environment can be classified into:

• interactions of the biomaterial on the biological environment manifested by: emission of metallic ions as a result of the chemical corrosion process (oxidation of metals), as well as the emission of metal particles caused by mechanical wear;

• interactions of the biological environment on the biomaterial through pH variation, caused by local inflammation and tissue destruction.

The most common form of tissue damage in the vicinity of Co-Cr-Ni alloy implants occurs in the form of granuloma, a phenomenon characterized by a high density of collagen fibers and the presence of giant multinucleated cells, MCG. The evolution of granulomas can sometimes lead to blockage of joint prostheses, which requires surgery on the implant. Another type of tissue reaction is manifested in the structural modification of the bone, in the implanted or prosthesed area. Thus, in the case of osteosynthesis with plates, the formation of new bone tissues was observed, after four years of exposure. Also there were found phenomena of osteolysis (bone destruction) usually produced in the case of uncimented implants, which cannot be rigorously fixed in the bone cavity and acquire a certain mobility over time.

6.4.4. Inflammatory reactions

Implants and metal prostheses can cause inflammatory reactions with the adjacent biological tissue, which manifests itself in different forms of inflammation of the cells [58]: histocytes, plasma cells, MGC, lymphocytes, etc.

6.4.5. Immunological reactions: Sensibility and allergy

Inflammatory and immunological symptoms are often very closely related to each other. Some metals such as iron, nickel and cobalt are known to have the ability to proliferate lymphocytes. Allergy is defined as a pronounced reaction of a substance when introduced into the body. A sensitive substance called an antigen is a molecule or a celua that, once introduced into the body, causes the formation of antibodies or defense cells.

Regarding the susceptibility to allergy, produced by the alloys used as surgical and orthopedic implants, the researchers' opinions are divided, but it is fully recognized that there is a relationship between the production of allergies and metal implants [59].

6.4.6. Carcinogenic effects

It has not been scientifically proven the existence of a direct relationship between the formation of cancer cells and the presence of implants in the human case. There have been reported cases of malignant fibrous histocytosis in the area of metal implants of plates and screws and other foreign bodies (shell fragments). There is, however, clear information on the long exposure of the human and animal bodies in environments containing metal compounds, in high concentrations, increases the risk of developing cancer. It is experimentally proven and recognized that metal powders released in the form of ions modify the biological metabolism and oxidation processes, nickel, cobalt and chromium compounds can have genotoxic and mutagenic effects. There are more and more frequent cases of malignant tumors in people in the area of long-use metal implants [60].

6.5. Specific technologies for obtaining ceramic coatings on metal biomaterials

The reason that leads to the increase in the use of ceramic coating technologies is their ability to improve the performance of some components or to modify the functional behavior of some materials. Using ceramic coatings, it is possible to make a compound that has performance of which, alone, the two components (neither coating nor substrate) are capable of [61, 62].

There are currently a multitude of coating techniques. They can generally be classified as atomic, participatory, comparative and surface modifier deposits.

There are also many other hybrid coating processes that have been developed to address certain specific needs.

The criteria that allow the evaluation of the effectiveness of each coating method are the following:

• during the coating process, the ceramic material does not alter irreversibly from a chemical and structural point of view;

• through the coating operation, the mechanical properties of the material that constitutes the implant support are not affected in an unfavourable way;

• the adhesion between the ceramic coating and the substrate must be large enough to be sure that, throughout the duration of use of the implant, there will be no interfacial damage between the materials brought into account;

• the additional cost of the coating process will not lead to a very large increase in the price of the implant [63-65].

Methods of realization of bioceramic coatings

In order to achieve bioceramic coatings, the following procedures and methods are used:

• thermal spray coating;

• sputtering;

• immersion-sintering coating (sol-gel method, immersion in melts);

• electrochemical coating (electrophoretic, electrolytic).

Themic spraying is typically separated into three major categories:

- combustive spraying (powder or electrode in flame, high-speed oxidizing fuel);

- electrode – arc (in air or inert gas);

- plasma (spraying in atmospheric plasma, in vacuum, in inert or inductive gas) [66-69].

Combustible spraying, it is the method of thermal spraying that uses mixtures of fuels with an oxidizing agent in order to produce a jet of hot gas that accelerates the particles injected into the staggered hot gases. The combustion spraying has been divided into two significant commercial technologies:

- spraying by combustion in the flame;

- high velocity oxy-fuel (HVDF) oxidizing combustion spraying.

Flame spraying is achieved by burning a mixture of gases at low pressure in the air at the exit of the burner. The materials are injected at the exit of the burner [70].

Flaccid spraying is generally limited to the deposition of materials with low melting points (bottles, enamels, SiO_2) because low combustion temperatures and the short residence time of particles in flame lead to low temperatures and velocities of the particles.

For higher melting temperatures and to increase the residence time in the flame (i.e., in the molten state) of the ceramic powder, the ROKIDE (Norton) process was developed. It uses ceramic bars obtained by sintering ceramic powders (including Al_2O_3 and ZrO_2). In this case, the ceramic bar is inserted into the center of the burner. As its end melts, the base is constantly pushed into the middle of the fire, where it melts, and a stream of air atomizes the molten material in droplets that are projected onto the surface of the substrate by means of the jet of gases.

The plasma spray coating is the only current technology that allows the realization of ceramic and metal deposits in thick layers (from 50 μm to several mm) on a multitude of supports that give them an improvement of their performance [71].

Plasma is the scientific name to describe the vapors of matter that possess a higher energy level than the usual gaseous state. Normal gases are composed of separate molecules. The plasma is composed of the same gas whose molecules have been dissociated in such a way that some of the electrically charged particles have been separated. By increasing the energy applied to these atoms, an ionization is obtained, the resulting gas being called plasma.

Atmospheric plasma is an environment obtained at temperatures higher than 3000°C; it consists of a large number of chemical species: ions, electrons, excited species, etc., obtained by dissociation of the molecules of a plasmagen gas, in an electric arc created between an anode and a cathode. Among the plasmagenic gases, the most commonly used, can be listed: pure argon, nitrogen, hydrogen, helium or mixtures thereof. This plasma provides a very intense caloric energy and a very strong kinetic energy.

The arousal energy necessary for plasma production is provided by a direct or alternating current electric generator, by radio frequency or microwave.

The material to be deposited, in the form of powder, injected into the carrier gas is melted and projected on a support with a speed that can reach 1800 m/sec.

The technology of thermal design in plasma is complex, taking into account the multitude of parameters involved in the process; the transfer of material on the support is carried out by storing (deposition) of particles. These powder particles are injected perpendicularly or countercurrently with the plasma jet at the outlet of the burner, by means of a carrier gas. They must be injected into the middle of the plasma jet before obtaining a total fusion (melting) of the powder. After a period of stationary of several milliseconds in the middle of the heat source, the particles are sprayed in the form of small droplets, which reach the impact with the surface of the substrate, forming the blades (layers) particularly fine.

The small drops of molten material are therefore subjected to a harsh thermal shock, thus being able to form an amorphous phase, as there is no time for recrystalization; the liquid state being suddenly frozen. Another consequence of this sudden decrease in temperature is the appearance of thermal voltages in the ceramic deposit, which makes the too thin coatings not have sufficient mechanical resistance, the induced microfisurge becoming very important.

The maintenance of the ceramic layer on the support is mainly determined by a mechanical connection with the roughness of the support surface or with the already solidified particles. In order to ensure the adhesion of the particles on the support, its surface is previously prepared by sandblasting and then flushed, in order to achieve a good rubble for adhesion and to avoid the presence of a film of impurities harmful to the adhesion [72].

In general, the ceramic deposits obtained in these conditions show a multitude of microcraps, which are form during the very rapid cold of the drops; at the same time, these deposits present a porous microstructure, which leads to an increase in the insulating power of the material.

Water-stabilised plasma has been in use for over 20 years, but only recently has it become commercially available. This process uses water as a plasma-forming medium. As with gas-stabilized systems, a cathode (typical of graphite) emits electrons, forming an electric arc that ionizes water, in the arc chamber, in O2- and H2+; thus appears a jet of high temperate temperature that is directed to an anode of Cu; ceramic particles are injected into this current.

The characteristics of the plasma essentially influence the quality of the deposited layer.

Currently, there are many types of plasma deposition facilities; the main ones being the following:

- spraying in atmospheric plasma;

- spraying in plasma in the atmosphere and controlled temeparture;

- spraying in plasma in a vacuum.

Of these, the deposits obtained by spraying in atmospheric plasma represent 98% of the total.

Two fundamental modes of operation are distinguished:

- with blown spring;

- with the transferred arc (continuous).

In a blown arc plasma, the plasma jet does not conduct current to the outside of the burner. This type of plasma is used to make ceramic deposits on metal or non-metallic substrates.

The regularity of feeding the ceramic powder in the plasma jet requires a rigorous particle size distribution of the plasma. Knowing the high viscosity of the plasma jet, a correct injection of the powder guarantees the obtaining of good quality deposits. The possibility of reaction between powder particles projected upon contact with ambient air is severely limited by the protective effect of the inert gas. Powders of highly reactive materials or of certain types of ceramics may undergo changes in chemical composition or variations in stoechiometry [73].

Cathor spray coating (sputtering), this is a technical by which a beam of ionized gas, with a very high speed, bombards a "target" (sample) of the material to be deposited, located in a vacuum chamber; very fine particles detached from the bombed material are deposited on a metal substrate placed in the path of the sprayed particles. In this way, a dense and adherent coating is forms on the substrate, by combining the positive effects of the high impact velocity and the presence of a reactive surface, without ions [74].

The method can also be used to cover metal aublayers with bioceramic materials, in order to obtain devices of medical interest.

The materials used as a substrate for the realization of coatings by this technical are: Ti, Ti alloys – 6Al- 4V, alloys of Type Co – Cr – Mo and alumina.

As a "target" or covering material, a disc made of sintered HA powders is used [75, 76].

As an ion source, gaseous argon is used, pumped into the vacuum chamber; it is ionized by means of electrons generated by a cathonous filament. The formed ions are focused by a magnetic field and accelerated by the shielding grids charged positively and negatively. A neutralization cathode that provides an additional number of electrons can also be used to neutralize the large energy beam that leave the ion source.

Before the actual coating is started, the ion beam is directed to the substrate sample to "spray" the impurities from their surface and the gases eventually absorbed on them. In this way, the superficial reactivity of the substrates is altered.

After the deposition, the analyses showed that it is generally amorphous or diffusely crystalline.

Therefore, in general, the coatings made by spraying the material by bombarding with gaseous ions, are subsequently subjected to heat treatments in order to improve the crystallity. A heat treatment at 500°C (for 30 - 60 min.) leads to obtaining crystalline and adherent coatings in the case of HA.

Immersion coating, two ways of achieving the coatings can be distinguished by this technical:

a) immersion coating – sintering;

b) immersion coating in melts.

a) The process of coating by immersion – sintering consists in immersing the substrate (either metallic or ceramic) in a bathroom containing a barbotin, with high viscosity, of HA powder.

The structures thus made are then sintered in a time-temperate cycle adequate to the densification of the ceramic coating of HA, e.g. At the temperature of 1100 - 1200°C, for a minimum of 3 hours.

b) The process of coating by immersion in melts consists in heating the bioceramic (for example HA) until melting and immersion of the metallic or ceramic substrate, for 3 to 5 seconds in the crucible containing the melt.

This immersion method is used to cover substrates made of Co-Cr–Mo alloy with a bioactive glass.

Ha melting at temperatures higher than 1500°C (in the alumina crucible in which titanium samples were immersed for 3 seconds) were also performed.

Electrochemical methods of coating: the coating by electrophoresis, a process similar to the immersion-sintering deposition is the electrophoretic coating, by which, suspensions or HA soils in electrolyte solutions, are deposited evenly on a metal layer, following the application of an electrical voltage from the outside; the coating thus made is then densified and connected to the metal substrate by sintering [77, 78].

By sintering at the temperature of 1000°C or at temperatures with higher value, high density HA coatings were obtained.

The adhesion of the deposit to the substrate is, however, below expectations, both in the case of sintering in the argon atmosphere and in the case of partial vacuum sintering (about 1 torr); in the latter case, the weak adhesion and the low strength of the connection between the substrate and the coating material are due to defects in the very thin interfacial oxide layer.

6.5. Methods to improve biocompatibility

The most used methods are aimed at improving the intrinsic biocompatibility, in the sense of inducing the bioactivity properties of titanium implants. They aim at speeding up the processes of adhesion, attachment, stretching and proliferation of osteoblast cells by applying thermal, chemical or different combinations of treatments thereof [80].

Upon closer examination of the interfaces between bone tissue and bioactive materials, it was discovered that most bioactive materials, with the exception of ceramics based on β-tricalcium phosphate and natural calcite, are connected to bone tissue by means of a layer of apatite. This layer can be reproduced on the surface of materials even in SBF (Simulated Body Fluid) with ionic concentrations approximately equal to those in human blood plasma. Detailed analyses of the superficial layer have shown that the formed apatite is similar in composition and structure to the bone mineral. Thus, osteogenic cells are expected to preferentially proliferate and differentiate to produce bone tissue on these surfaces, just as they do with fractured bone. As a result, new bone tissue can develop from the environment adjacent to the implant to come into direct contact with the apatitis on the surface of the implant. When this phenomenon occurs, a close chemical bond is made between the bone mineral and the apatite on the surface of the implant [81].

6.5.1. Chemical treatments

Treatments in alkaline environment

Due to the very high affinity, it has towards oxygen, implants made of titanium are covered by a passive layer of oxide, which makes the material bioinert, that is, not to react in any way with adjacent tissues. To create this layer, the material can be subjected to treatment in alkaline medium by immersion in a hard base solution. The most widely used base is NaOH.

After chemical treatment in the alkaline medium, on the surface of pure commercial titanium, a large amount of calcium phosphates and calcium silico-titanate are deposited. These, together with the oh$^-$ active groupings contribute to the formation of a micron-level roughness that improves the adhesion capacity of the cells or the bone mineral to the surface of the implant. Following contact with the biological environment, sodium ions are replaced by calcium ions and a soluble layer is formed on which hydroxyapatite is to be nucleus [82, 83].

Treatments in the acidic environment

Treatment in the acidic environment involves immersion in solutions of HCl, HF, H_3PO_4, H_2SO_4 or HNO_3 preceded by pickling. The effect of the treatment is more intense if it is carried out electrolytely or is followed by a less aggressive basic treatment than the usual one.

6.5.2. Ionic implantation in plasma

The process of modifying the surface by ionic implantation is based on the bombardment of the surface layer of the material with ions charged with very high energies.

This process aims to improve the corrosion and wear resistance properties by forming on the surface a thin film that has tribological, anticorrosive and mechanical properties superior to the internal structure [84].

Deposits of oxides modified from titanium by electrochemical processes: one of the methods of deposition of hydroxyapatite layers is electrochemical deposition, a process that is especially addressed to irregularly shaped parts. The process is carried out in a relatively short time and at low temperatures. The parameters of the newly deposited layer such as thickness and composition can be easily controlled by electrodeposition processes [85].

6.5.3. Other methods to improve intrinsic biocompatibility

Bioactive layer deposition

Deposition of organic molecules - By this procedure, substances are deposited on the surface of the material that have the role of being the target for integrins in the process of cell attachment.

Sol-gel coatings – the sol-gel technique is an alternative for deposition of bioactive layers, with advantages such as advanced purity and chemical homogeneity, fine structure, low processing temperature and controllable layer thickness [86-88].

Cathodic deposition – the biocompatibility of titanium can also be achieved by a method of cathodic deposition of solutions containing calcium phosphates. The aim is to maintain in the long term the fixation capacity of the implant in the adjacent bone tissue, through the continuous development of a permanent interface [89, 90].

References

[1] Corbett, R.A. Laboratory Corrosion Testing of Medical Implants; Corrosion Testing Laboratories, Inc.: Newark, DE, USA, **2004**. https://doi.org/10.1515/CORRREV.2003.21.2-3.231

[2] Kamachi Mudali, U.; Sridhar, T.M.; Eliaz, N.; Raj, B. Failures of stainless steel orthopaedic devices—causes and remedies. *Corros. Rev.* **2003**, *21*, 231–267.

[3] Hiromoto, S.; Hanawa, T. Corrosion of implant metals in the presence of cells. *Corros. Rev.* **2006**, *24*, 323–352. https://doi.org/10.1515/CORRREV.2006.24.5-6.323

[4] Zitter, H. Case histories on surgical implants and their causes. *Werkstoffe Korrosion* **1992**, *42*, 455–466. https://doi.org/10.1002/maco.19910420904

[5] Ryhänen, J. Biocompatibility Evaluation of Nickel-Titanium Shape Memory Metal Alloy. Ph.D. Thesis, University of Oulu, Oulu, Finland, **1999**.

[6] Morita, M.; Sasada, T.; Hayashi, H.; Tsukamoto, Y. The corrosion fatigue properties of surgical implants in a living body. *J. Biomed. Mater. Res.* **1988**, *22*, 529–540. https://doi.org/10.1002/jbm.820220608

[7] Hanawa, T. Reconstruction and regeneration of surface oxide film on metallic materials in biological environments. *Corros. Rev.* **2003**, *21*, 161–182. https://doi.org/10.1515/CORRREV.2003.21.2-3.161

[8] Von Fraunhofer, J.A.; Berberich, N.; Seligson, D. Antibiotic-metal interactions in saline medium. *Biomaterials* **1989**, *10*, 136–138. https://doi.org/10.1016/0142-9612(89)90048-3

[9] Bălțatu, M.S.; Vizureanu, P.; Mareci, D.; Burtan, L.C.; Chiruță, C.; L.C Trincă, Effect of Ta on the electrochemical behavior of new TiMoZrTa alloys in artificial physiological solution simulating in vitro inflammatory conditions. *Materials and Corrosion* **2016**, 67(12), 1314-1320. https://doi.org/10.1002/maco.201609041

[10]   Cumpătă, C.N. Implaturi acoperite chimic cu hidroxiapatita biologică, *Ed. Printech*, București, **2012**.

[11]   Ludwigson, D.C. Requirements for metallic surgical implants and prosthetic devices. *Metals Engineering Quarterly*: American Society of Metallurgists 1, **1965**.

[12]   Owens, G.J.; Singh, R.K.; Foroutan, F.; Alqaysi, M.; Han, C.-M.; Mahapatra, C.; Kim, H.-W.; Knowles, J.C. Sol–gel based materials for biomedical applications. *Prog. Mater. Sci.* **2016**, *77*, 1–79. https://doi.org/10.1016/j.pmatsci.2015.12.001

[13]   Prabhu, S.; Poulose, E.K. Silver nanoparticles: Mechanism of antimicrobial action, synthesis, medical applications, and toxicity effects. *Int. Nano Lett.*   **2012**, *2*, 32. https://doi.org/10.1186/2228-5326-2-32

[14]   Perez, R.A.; Won, J.E.; Knowles, J.C.; Kim, H.W. Naturally and synthetic smart composite biomaterials for tissue regeneration. *Adv. Drug Deliv. Rev.* **2013**, *65*, 471–496. https://doi.org/10.1016/j.addr.2012.03.009

[15]   Mohseni, E.; Zalnezhad, E.; Bushroa, A.R. Comparative investigation on the adhesion of hydroxyapatite coating on Ti–6Al–4V implant: A review paper. *Int. J. Adhes. Adhes.* **2014**, *48*, 238–257. https://doi.org/10.1016/j.ijadhadh.2013.09.030

[16]   Trivedi, P.; Gupta, P.; Srivastava, S.; Jayaganthan, R.; Chandra, R.; Roy, P. Characterization and in vitro biocompatibility study of Ti–Si–N nanocomposite coatings developed by using physical vapor deposition. *Appl. Surf. Sci.* **2014**,   *293*, 143–150. https://doi.org/10.1016/j.apsusc.2013.12.119

[17]   Long, Y.; Javed, A.; Chen, J.; Chen, Z.-k.; Xiong, X. Phase composition, microstructure and mechanical properties of ZrC coatings produced by chemical vapor deposition. *Ceram. Int.* **2014**, *40*, 707–713. https://doi.org/10.1016/j.ceramint.2013.06.059

[18]   Das, K.; Bose, S.; Bandyopadhyay, A.; Karandikar, B.; Gibbins, B.L. Surface coatings for improvement of bone cell materials and antimicrobial activities of Ti

implants. *J. Biomed. Mater. Res. B Appl. Biomater.* **2008**, *87*, 455–460. https://doi.org/10.1002/jbm.b.31125

[19] Hu, J.; Zhong, X.; Fu, X. Enhanced bone remodeling effects of low-modulus Ti-5Zr-3Sn-5Mo-25Nb alloy implanted in the mandible of beagle dogs under delayed loading. *ACS Omega* **2019**, *4*, 18653–18662. https://doi.org/10.1021/acsomega.9b02580

[20] Wang, Y.; Papadimitrakopoulos, F.; Burgess, D.J. Polymeric "smart" coatings to prevent foreign body response to implantable biosensors. *J. Control. Release* ***2013***, *169*, 341–347. https://doi.org/10.1016/j.jconrel.2012.12.028

[21] Campoccia, D.; Montanaro, L.; Arciola, C.R. A review of the biomaterials technologies for infection-resistant surfaces. *Biomaterials* **2013**, *34*, 8533–8554. https://doi.org/10.1016/j.biomaterials.2013.07.089

[22] Getzlaf, M.A.; Lewallen, E.A.; Kremers, H.M.; Jones, D.L.; Bonin, C.A.; Dudakovic, A.; Thaler, R.; Cohen, R.C.; Lewallen, D.G.; van Wijnen, A.J. Multi-disciplinary antimicrobial strategies for improving orthopaedic implants to prevent prosthetic joint infections in hip and knee. *J. Orthop. Res.* **2016**, *34*, 177–186. https://doi.org/10.1002/jor.23068

[23] Goodman, S.B.; Yao, Z.; Keeney, M.; Yang, F. The future of biologic coatings for orthopaedic implants. *Biomaterials* **2013**, *34*, 3174–3183. https://doi.org/10.1016/j.biomaterials.2013.01.074

[24] Brammer, K.S.; Oh, S.; Cobb, C.J.; Bjursten, L.M.; van der Heyde, H.; Jin, S. Improved bone-forming functionality on diameter-controlled TiO(2) nanotube surface. *Acta Biomater.* **2009**, *5*, 3215–3223. https://doi.org/10.1016/j.actbio.2009.05.008

[25] D. Mareci, R. Chelariu, D.M. Gordin, G. Ungureanu, T. Gloriant, Comparative corrosion study of Ti–Ta alloys for dental applications, *Acta Biomaterialia* **2009**, *5,* 3625–3639. https://doi.org/10.1016/j.actbio.2009.05.037

[26] J.R.S. Junior, R.A. Nogueira, R. Oliveira de Araújo, T.A.G. Donato, V.E.A. Chavez, A.P.R.A. Claro, J.C.S.M. Moraes, M.A.R. Buzalaf, C.R. Grandini, Preparation and Characterization of Ti-15Mo Alloy used as Biomaterial, *Materials Research* **2011**, *14(1)*, 107–112. https://doi.org/10.1590/S1516-14392011005000013

[27] Shih, C.C.; Lin, S.J.; Chung, K.H.; Chen, Y.L.; Su, Y.Y. Increased corrosion resistance of stent materials by converting current surface film of polycrystalline

oxide into amorphous oxide. *J. Biomed. Mater. Res.* **2000**, *52*, 323–332.
https://doi.org/10.1002/1097-4636(200011)52:2<323::AID-JBM11>3.0.CO;2-Z

[28]   D.R.N. Correa, F.B. Vicente, T.A.G. Donato, V.E. Arana-Chavez, M.A.R.
Buzalaf, C.R. Grandini, The effect of the solute on the structure, selected mechanical
properties, and biocompatibility of Ti–Zr system alloys for dental applications,
Materials Science and Engineering C, **2014**, *34*, 354–359.
https://doi.org/10.1016/j.msec.2013.09.032

[29]   Elsner, J.J.; Eliaz, N.; Linder-Ganz, E. The use of polyurethanes in joint
replacement. In Materials for Joint Arthroplasty: Biotribology of Potential Bearings;
Sonntag, R., Kretzer, J.P., Eds.; *Imperial College Press: London*, UK, **2016**, 259–298.
https://doi.org/10.1142/9781783267170_0009

[30]   Standard Test Method for Conducting Cyclic Potentiodynamic Polarization
Measurements for Localized Corrosion Susceptibility of Iron-, Nickel-, or Cobalt-
Based Alloys; ASTM G61–86(2018); *ASTM International: West Conshohocken*, PA,
USA, **2018.**

[31]   Standard Specification for Wrought Nitrogen Strengthened 21Chromium-
10Nickel-3Manganese-2.5Molybdenum Stainless Steel Alloy Bar for Surgical
Implants (UNS S31675); ASTM F1586–13e1; *ASTM: West Conshohocken*, PA, USA,
2013.

[32]   Standard Specification for Wrought Stainless Steels for Surgical Instruments;
ASTM F899–12b; *ASTM: West Conshohocken*, PA, USA, **2012.**

[33]   Standard Test Method for Corrosion of Surgical Instruments; ASTM F1089–18;
ASTM: West Conshohocken, PA, USA, **2018.**

[34]   Standard Specification for Wrought 18Chromium-14Nickel-2.5Molybdenum
Stainless Steel Bar and Wire for Surgical Implants (UNS S31673); ASTM F138–13a;
ASTM: West Conshohocken, PA, USA, **2013.**

[35]   Elsner, J.J.; Shemesh, M.; Mezape, Y.; Levenshtein, M.; Hakshur, K.; Shterling,
A.; Linder-Ganz, E.; Eliaz, N. Long-term evaluation of a compliant cushion form
acetabular bearing for hip joint replacement: A 20 million cycles wear simulation. *J.
Orthop. Res.* **2011**, *29*, 1859–1866. https://doi.org/10.1002/jor.21471

[36]   Mendel, K.; Eliaz, N.; Benhar, I.; Hendel, D.; Halperin, N. Magnetic isolation of
particles suspended in synovial fluid for diagnostics of natural joint chondropathies.
Acta Biomater. **2010**, *6*, 4430–4438. https://doi.org/10.1016/j.actbio.2010.06.003

[37]   Meyer, D.M.; Tillinghast, A.; Hanumara, N.C.; Franco, A. Bio-ferrography to capture and separate polyethylene wear debris from hip simulator fluid and compared with conventional filter method. *J. Tribol.* **2006**, *128*, 436–441. https://doi.org/10.1115/1.2162554

[38]   Standard Practice for Calculation of Corrosion Rates and Related Information from Electrochemical Measurements; ASTM G102–89, ASTM International: West Conshohocken, PA, USA, **2015**.

[39]   Ifergane, S.; Eliaz, N.; Stern, N.; Kogan, E.; Shemesh, G.; Sheinkopf, H.; Eliezer, D. The effect of manufacturing processes on the fatigue lifetime of aeronautical bolts. *Eng. Fail. Anal.* **2001**, *8*, 227–235. https://doi.org/10.1016/S1350-6307(00)00013-3

[40]   M.S. Bălțatu, P. Vizureanu, M. Benchea, M.G. Minciună, A.C. Achiței, B. Istrate, Ti-Mo-Zr-Ta Alloy for Biomedical Applications: Microstructures and Mechanical Properties, *Key Engineering Materials*, **2018**, *750*, 184-188. https://doi.org/10.4028/www.scientific.net/KEM.750.184

[41]   Mor, G.K.; Varghese, O.K.; Paulose, M.; Shankar, K.; Grimes, C.A. A review on highly ordered, vertically oriented TiO_2 nanotube arrays: Fabrication, material properties, and solar energy applications. *Sol. Energy Mater. Sol. Cells* **2006**, *90*, 2011–2075. https://doi.org/10.1016/j.solmat.2006.04.007

[42]   Puppi, D.; Chiellini, F.; Piras, A.M.; Chiellini, E. Polymeric materials for bone and cartilage repair. *Prog. Polym. Sci.* **2010**, *35*, 403–440. https://doi.org/10.1016/j.progpolymsci.2010.01.006

[43]   Gallo, J.; Holinka, M.; Moucha, C.S. Antibacterial surface treatment for orthopaedic implants. *Int. J. Mol. Sci.* **2014**, *15*, 13849–13880. https://doi.org/10.3390/ijms150813849

[44]   Hakshur, K.; Benhar, I.; Bar-Ziv, Y.; Halperin, N.; Segal, D.; Eliaz, N. The effect of hyaluronan injections into human knees on the number of bone and cartilage war particles captured by bio-ferrography. *Acta Biomater.* **2011**, *7*, 848–857. https://doi.org/10.1016/j.actbio.2010.08.030

[45]   Geetha, M.; Singh, A.K.; Asokamani, R.; Gogia, A.K. Ti based biomaterials, the ultimate choice for orthopaedic implants—A review. *Prog. Mater. Sci.* **2009**, *54*, 397–425. https://doi.org/10.1016/j.pmatsci.2008.06.004

[46]   Asri, R.I.; Harun, W.S.; Hassan, M.A.; Ghani, S.A.; Buyong, Z. A review of hydroxyapatite-based coating techniques: Sol-gel and electrochemical depositions on

biocompatible metals. *J. Mech. Behav. Biomed. Mater.* **2016**, *57*, 95–108.
https://doi.org/10.1016/j.jmbbm.2015.11.031

[47]　A.C. Bărbînță, Îmbunătățirea proprietăților aliajelor de Ti-Nb-Zr-Ta utilizate la fabricarea protezelor ortopedice - teză de doctorat, Iași, **2003**.

[48]　F. Miculescu, Tehnici de analiză și control a biomaterialelor, *Ed. Printech*, București, **2009**.

[49]　Hallab, N.J. Hypersensitivity to implant debris. In Degradation of Implant Materials; Eliaz, N., *Ed.; Springer: New York*, NY, USA, **2012**, *12*, 329–345. https://doi.org/10.1007/978-1-4614-3942-4_12

[50]　Aksakal, B.; Yildirim, Ö.S.; Gul, H. Metallurgical failure analysis of various implant materials used in orthopedic applications. *J. Fail. Anal. Prev.* **2004**, *4*, 17–23. https://doi.org/10.1007/s11668-996-0007-9

[51]　Urban, R.M.; Jacobs, J.J.; Gilbert, J.L.; Galante, J.O. Migration of corrosion products from modular hip prostheses. Particle microanalysis and histopathological findings. *J. Bone Joint Surg. Am.* **1994**, *76*, 1345–1359. https://doi.org/10.2106/00004623-199409000-00009

[52]　Traisnel, M.; Le Maguer, D.; Hildebrand, H.F.; Iost, A. Corrosion of surgical implants. *Clin. Mater.* **1990**, *5*, 309–318. https://doi.org/10.1016/0267-6605(90)90030-Y

[53]　Russell, A.P.; Westcott, V.C.; Demaria, A.; Johns, M. The concentration and separation of bacteria and cells by ferrography. *Wear* **1983**, *90*, 159–165. https://doi.org/10.1016/0043-1648(83)90054-6

[54]　Eliaz, N.; Gileadi, E. Physical Electrochemistry: Fundamentals, Techniques, and Applications, 2nd ed.; *Wiley-VCH: Weinheim*, Germany, **2019**; ISBN 978-3-527-34139-9.

[55]　Bălțatu, M.S.; Vizureanu, P.; Cimpoeșu, R.; Abdullah, M.M.A.B.; Sandu, A.V. The Corrosion Behavior of TiMoZrTa Alloys Used for Medical Applications, *Revista de Chimie* **2016**, *67(10)*, 2100-2002.

[56]　Huang, Y.; Han, S.; Pang, X.; Ding, Q.; Yan, Y. Electrodeposition of porous hydroxyapatite/calcium silicate composite coating on titanium for biomedical applications. *Appl. Surf. Sci.* **2013**, *271*, 299–302. https://doi.org/10.1016/j.apsusc.2013.01.187

[57]　Mahadik, D.B.; Lakshmi, R.V.; Barshilia, H.C. High performance single layer nano-porous antireflection coatings on glass by sol–gel process for solar energy

applications. Sol. Energy Mater. *Sol. Cells* **2015**, *140*, 61–68.
https://doi.org/10.1016/j.solmat.2015.03.023

[58]   Surmenev, R.A.; Surmeneva, M.A.; Ivanova, A.A. Significance of calcium
phosphate coatings for the enhancement of new bone osteogenesis–a review. *Acta
Biomater.* **2014**, *10*, 557–579. https://doi.org/10.1016/j.actbio.2013.10.036

[59]   Fernandes, E.M.; Pires, R.A.; Mano, J.F.; Reis, R.L. Bionanocomposites from
lignocellulosic resources: Properties, applications and future trends for their use in the
biomedical field. *Prog. Polym. Sci.* **2013**, *38*, 1415–1441.
https://doi.org/10.1016/j.progpolymsci.2013.05.013

[60]   Catauro, M.; Bollino, F.; Veronesi, P.; Lamanna, G. Influence of PCL on
mechanical properties and bioactivity of ZrO2-based hybrid coatings synthesized by
sol-gel dip coating technique. *Mater. Sci. Eng. C Mater. Biol. Appl.* **2014**, *39*, 344–
351. https://doi.org/10.1016/j.msec.2014.03.025

[61]   Gopi, D.; Collins Arun Prakash, V.; Kavitha, L.; Kannan, S.; Bhalaji, P.R.;
Shinyjoy, E.; Ferreira, J.M.F. A facile electrodeposition of hydroxyapatite onto borate
passivated surgical grade stainless steel. *Corros. Sci.* **2011**, *53*, 2328–2334.
https://doi.org/10.1016/j.corsci.2011.03.018

[62]   Türk, S.; Altınsoy, `I.; ÇelebiEfe, G.; Ipek, M.; Özacar, M.; Bindal, C.
Microwave–assisted biomimetic synthesis of hydroxyapatite using different sources
of calcium. *Mater. Sci. Eng. C* **2017**, *76*, 528–535.
https://doi.org/10.1016/j.msec.2017.03.116

[63]   Lugovskoy, A.; Lugovskoy, S. Production of hydroxyapatite layers on the plasma
electrolytically oxidized surface of titanium alloys. *Mater. Sci. Eng. C Mater. Biol.
Appl.* **2014**, *43*, 527–532. https://doi.org/10.1016/j.msec.2014.07.030

[64]   Yang, Y.; Cheng, Y.F. Fabrication of Ni–Co–SiC composite coatings by pulse
electrodeposition—Effects of duty cycle and pulse frequency. *Surf. Coat. Technol.*
2013, *216*, 282–288. https://doi.org/10.1016/j.surfcoat.2012.11.059

[65]   Lim, C.S. Upconversion photoluminescence properties of
SrY2(MoO4)4:Er3+/Yb3+ phosphors synthesized by a cyclic microwave-modified
sol–gel method. *Infrared Phys. Technol.* **2014**, *67*, 371–376.
https://doi.org/10.1016/j.infrared.2014.08.018

[66]   Liu, C.; Dong, J.Y.; Yue, L.L.; Liu, S.H.; Wan, Y.; Liu, H.; Tan, W.Y.; Guo, Q.Q.;
Zhang, D. Rapamycin/sodium hyaluronate binding on nano-hydroxyapatite coated

titanium surface improves MC3T3-E1 osteogenesis. *PLoS ONE* **2017**, *12*, e0171693. https://doi.org/10.1371/journal.pone.0171693

[67]   Ching, H.A.; Choudhury, D.; Nine, M.J.; Abu Osman, N.A. Effects of surface coating on reducing friction and wear of orthopaedic implants. *Sci. Technol. Adv. Mater.* **2014**, *15*, 014402. https://doi.org/10.1088/1468-6996/15/1/014402

[68]   Chen, Q.; Cabanas-Polo, S.; Goudouri, O.M.; Boccaccini, A.R. Electrophoretic co-deposition of polyvinyl alcohol (PVA) reinforced alginate-Bioglass(R) composite coating on stainless steel: Mechanical properties and in-vitro bioactivity assessment. *Mater. Sci. Eng. C Mater. Biol. Appl.* **2014**, *40*, 55–64. https://doi.org/10.1016/j.msec.2014.03.019

[69]   Pishbin, F.; Mourino, V.; Flor, S.; Kreppel, S.; Salih, V.; Ryan, M.P.; Boccaccini, A.R. Electrophoretic deposition of gentamicin-loaded bioactive glass/chitosan composite coatings for orthopaedic implants. *ACS Appl. Mater. Interfaces* **2014**, *6*, 8796–8806. https://doi.org/10.1021/am5014166

[70]   Calderón, J.A.; Henao, J.E.; Gómez, M.A. Erosion–corrosion resistance of Ni composite coatings with embedded SiC nanoparticles. Electrochim. *Acta* **2014**, *124*, 190–198. https://doi.org/10.1016/j.electacta.2013.08.185

[71]   Min, J.; Braatz, R.D.; Hammond, P.T. Tunable staged release of therapeutics from layer-by-layer coatings with clay interlayer barrier. *Biomaterials* **2014**, *35*, 2507–2517. https://doi.org/10.1016/j.biomaterials.2013.12.009

[72]   Montemor, M.F. Functional and smart coatings for corrosion protection: A review of recent advances. *Surf. Coat. Technol.* **2014**, *258*, 17–37. https://doi.org/10.1016/j.surfcoat.2014.06.031

[73]   Boke, F.; Giner, I.; Keller, A.; Grundmeier, G.; Fischer, H. Plasma-enhanced chemical vapor deposition (PE-CVD) yields better hydrolytical stability of biocompatible SiOx thin films on implant alumina ceramics compared to rapid thermal evaporation Physical Vapor Deposition (PVD). *ACS Appl. Mater. Interfaces* **2016**, *8*, 17805–17816. https://doi.org/10.1021/acsami.6b04421

[74]   Raphel, J.; Holodniy, M.; Goodman, S.B.; Heilshorn, S.C. Multifunctional coatings to simultaneously promote osseointegration and prevent infection of orthopaedic implants. *Biomaterials* **2016**, *84*, 301–314. https://doi.org/10.1016/j.biomaterials.2016.01.016

[75]　Ciobanu, G.; Ciobanu, O. Investigation on the effect of collagen and vitamins on biomimetic hydroxyapatite coating formation on titanium surfaces. *Mater. Sci. Eng. C Mater. Biol. Appl.* **2013**, *33*, 1683–1688. https://doi.org/10.1016/j.msec.2012.12.080

[76]　Qureshi, S.; Zheng, Z.; Sarwar, M.; Félix, O.; Decher, G. Nanoprotective layer-by-layer coatings with epoxy components for enhancing abrasion resistance: Toward robust multimaterial nanoscale films. *ACS Nano* **2013**, *7*, 9336–9344. https://doi.org/10.1021/nn4040298

[77]　Xie, Y.; Li, H.; Zhang, C.; Gu, X.; Zheng, X.; Huang, L. Graphene-reinforced calcium silicate coatings for load-bearing implants. *Biomed. Mater.* **2014**, *9*, 025009. https://doi.org/10.1088/1748-6041/9/2/025009

[78]　Bakhsheshi-Rad, H.R.; Hamzah, E.; Daroonparvar, M.; Saud, S.N.; Abdul-kadir, M.R. Bi-layer nano-TiO2/FHA composite coatings on Mg–Zn–Ce alloy prepared by combined physical vapour deposition and electrochemical deposition methods. *Vacuum* **2014**, *110*, 127–135. https://doi.org/10.1016/j.vacuum.2014.08.013

[79]　Bagherifard, S. Mediating bone regeneration by means of drug eluting implants: From passive to smart strategies. *Mater. Sci. Eng. C Mater. Biol. Appl.* **2017**, *71*, 1241–1252. https://doi.org/10.1016/j.msec.2016.11.011

[80]　Xin, W.; Meng, C.; Jie, W.; Junchao, T.; Yan, S.; Ning, D. Morphology dependence of TiO2nanotube arrays on anodization variables and buffer medium. *J. Semicond.* **2010**, *31*, 063003. https://doi.org/10.1088/1674-4926/31/6/063003

[81]　Huang, Y.; Yan, Y.; Pang, X.; Ding, Q.; Han, S. Bioactivity and corrosion properties of gelatin-containing and strontium-doped calcium phosphate composite coating. *Appl. Surf. Sci.* **2013**, *282*, 583–589. https://doi.org/10.1016/j.apsusc.2013.06.015

[82]　Liu, K.; Tian, Y.; Jiang, L. Bio-inspired superoleophobic and smart materials: Design, fabrication, and application. *Prog. Mater. Sci.* **2013**, *58*, 503–564. https://doi.org/10.1016/j.pmatsci.2012.11.001

[83]　Kılıç, F.; Gül, H.; Aslan, S.; Alp, A.; Akbulut, H. Effect of CTAB concentration in the electrolyte on the tribological properties of nanoparticle SiC reinforced Ni metal matrix composite (MMC) coatings produced by electrodeposition. *Colloids Surf. A Physicochem. Eng. Asp.* **2013**, *419*, 53–60. https://doi.org/10.1016/j.colsurfa.2012.11.048

[84]　Arafat, M.T.; Lam, C.X.; Ekaputra, A.K.; Wong, S.Y.; Li, X.; Gibson, I. Biomimetic composite coating on rapid prototyped scaffolds for bone tissue

engineering. *Acta Biomater.* **2011**, *7*, 809–820.
https://doi.org/10.1016/j.actbio.2010.09.010

[85] Sharifi, E.; Azami, M.; Kajbafzadeh, A.M.; Moztarzadeh, F.; Faridi-Majidi, R.; Shamousi, A.; Karimi, R.; Ai, J. Preparation of a biomimetic composite scaffold from gelatin/collagen and bioactive glass fibers for bone tissue engineering. Mater. Sci. Eng. *C Mater. Biol. Appl.* **2016**, *59*, 533–541.
https://doi.org/10.1016/j.msec.2015.09.037

[86] Xing, R.; Jiao, T.; Yan, L.; Ma, G.; Liu, L.; Dai, L.; Li, J.; Möhwald, H.; Yan, X. Colloidal gold–collagen protein core–shell nanoconjugate: One-step biomimetic synthesis, layer-by-layer assembled film, and controlled cell growth. ACS Appl. Mater. *Interfaces* **2015**, *7*, 24733–24740. https://doi.org/10.1021/acsami.5b07453

[87] Benea, L.; Mardare-Danaila, E.; Mardare, M.; Celis, J.-P. Preparation of titanium oxide and hydroxyapatite on Ti–6Al–4V alloy surface and electrochemical behaviour in bio-simulated fluid solution. *Corros. Sci.* **2014**, *80*, 331–338.
https://doi.org/10.1016/j.corsci.2013.11.059

[88] Wu, S.; Liu, X.; Yeung, K.W.K.; Liu, C.; Yang, X. Biomimetic porous scaffolds for bone tissue engineering. *Mater. Sci. Eng. R Rep.* **2014**, *80*, 1–36.
https://doi.org/10.1016/j.mser.2014.04.001

[89] Xia, F.; Xu, H.; Liu, C.; Wang, J.; Ding, J.; Ma, C. Microstructures of Ni–AlN composite coatings prepared by pulse electrodeposition technology. *Appl. Surf. Sci.* **2013**, *271*, 7–11. https://doi.org/10.1016/j.apsusc.2012.12.064

[90] Wang, Z.; Zhang, X.; Gu, J.; Yang, H.; Nie, J.; Ma, G. Electrodeposition of alginate/chitosan layer-by-layer composite coatings on titanium substrates. *Carbohydr. Polym.* **2014**, *103*, 38–45. https://doi.org/10.1016/j.carbpol.2013.12.007

About the Authors

Madalina-Simona BALTATU.
Lecturer Ph.D. Eng.
Department of Technology and Equipment for Materials Processing, Faculty of Materials Science and Engineering, "Gheorghe Asachi" Technical University of Iasi
cercel.msimona@yahoo.com/ http://www.afir.org.ro/msb/.

Gained her PhD Eng. in 2017. She is focused on developing new biomaterials and advanced characterization. She published over 30 articles, 2 books, one international book chapter, 5 patent applications and she carry out activities in 6 projects.

Dumitru-Doru BURDUHOS-NERGIS
Assistant Professor PhD.Eng.
Department of Technology and Equipment for Materials Processing, Faculty of Materials Science and Engineering, "Gheorghe Asachi" Technical University of Iasi
doru.burduhos@tuiasi.ro/ https://www.afir.org.ro/ddbn/.

Materials engineering researcher with 5 years of experience in the field of geopolymers. The research activity in the field carried out during the elaboration of the thesis for the master's degree graduation, was continued within the PhD stage, starting in 2017, and the scientific research results were disseminated 24 publications, of which 7 articles were published in Web of Science (WoS) (3 as the first author), 4 articles published in proceedings, indexed in WoS, 6 in conference proceedings indexed in SCOPUS, 5 articles published in BDI-listed journals, two international books and two international book chapters

Diana Petronela BURDUHOS-NERGIS
Assistant Professor PhD.Eng.
"Gheorghe Asachi" Technical University of Iasi
diana.burduhos@tuiasi.ro, www.afir.org.ro/dpbn

Researcher and assistant professor at Gheorghe Asachi Technical University of Iasi, Faculty of Materials Science and Engineering, with a doctoral thesis on the study and improvement of carbon steel components in personal protective equipment by depositing different types of coatings, being involved in scientific research since she was a student. She has over 18 publication, 14 of them indexed by Web of Science. She has many awards received from presentations at conferences or invention exhibitions.

Advanced Metallic Biomaterials

Materials Research Forum LLC

Materials Research Foundations **118** (2022)

https://doi.org/10.21741/9781644901779

Petrica VIZUREANU

Professor Ph.D. Eng.

Head of department at Department of Technology and Equipment for Materials Processing, Faculty of Materials Science and Engineering, "Gheorghe Asachi" Technical University of Iasi

peviz2002@yahoo.com/ http://afir.org.ro/peviz/

Professor and researcher at "Gheorghe Asachi" Technical University of Iasi, with more than 30 years of experience. Ph.D. degree, since 1999 in Materials science and engineering; 2010 - present Ph.D. Supervisor in Materials Engineering domain. He has over 150 publications, 130 articles being indexed in Web of Science. He has vast experience in the field of composite materials; ceramic materials, insulating materials; optimization of materials characteristics. H-index is 18.

www.ingramcontent.com/pod-product-compliance
Lightning Source LLC
Chambersburg PA
CBHW071647210326
41597CB00017B/2140